T0213991

SpringerBriefs in Statistics

SpringerBriefs present concise summaries of cutting-edge research and practical applications across a wide spectrum of fields. Featuring compact volumes of 50 to 125 pages, the series covers a range of content from professional to academic. Typical topics might include:

- A timely report of state-of-the art analytical techniques
- A bridge between new research results, as published in journal articles, and a contextual literature review
- A snapshot of a hot or emerging topic
- An in-depth case study or clinical example
- A presentation of core concepts that students must understand in order to make independent contributions

SpringerBriefs in Statistics showcase emerging theory, empirical research, and practical application in Statistics from a global author community.

SpringerBriefs are characterized by fast, global electronic dissemination, standard publishing contracts, standardized manuscript preparation and formatting guidelines, and expedited production schedules.

More information about this series at http://www.springer.com/series/8921

Tony Pourmohamad • Herbert K. H. Lee

Bayesian Optimization with Application to Computer Experiments

 Springer

Tony Pourmohamad
Genentech
South San Francisco, CA, USA

Herbert K. H. Lee
Department of Statistics,
Baskin School of Engineering
University of California, Santa Cruz
Santa Cruz, CA, USA

ISSN 2191-544X   ISSN 2191-5458   (electronic)
SpringerBriefs in Statistics
ISBN 978-3-030-82457-0   ISBN 978-3-030-82458-7   (eBook)
https://doi.org/10.1007/978-3-030-82458-7

This Springer imprint is published by the registered company Springer Nature Switzerland AG
The registered company address is: Gewerbestrasse 11, 6330 Cham, Switzerland

To Josie, Vincent, Ava, and Spike

Preface

Optimization is a field with a long history and much past research. Our focus here is on a very specific sub-area, that of derivative-free optimization of expensive black-box functions. While much of the optimization world involves efficient algorithms that make use of gradient information, we operate in an environment where gradient information is typically not available. The application area we are most familiar with is that of deterministic computer simulation experiments, where complex computer code attempts to model a real-world phenomenon. The code can be run for any inputs, but it only returns output values, and does not provide any gradient information for those outputs. Also, running the code for the same input will always return the same output. The code can be expensive to run, requiring significant computing power and time. Thus, there is a need for efficient optimization routines that do not rely on information about derivatives of the function. That is the focus of this book.

Moreover, there are few books that are devoted to either the topic of Bayesian optimization or computer experiments, and even fewer devoted solely to both. The term "Bayesian optimization" comes from the machine learning literature and has become a very hot topic among the machine learning community due to its highly successful application in the optimization of tuning parameters for machine learning models. Much work has also been done in the realms of applied math, statistics, and engineering, sometimes under different names. The mixing of these disciplines has led to useful insights and new research directions. And although the roots of Bayesian optimization can be traced back to early optimization problems in computer experiments, the application of Bayesian optimization is still in its nascent phase within the field of computer experiments as compared to in machine learning.

This book is intended to provide a quick introduction to the topic of Bayesian optimization, with aim to introduce the reader to Bayesian optimization, highlight advances in Bayesian optimization, and showcase its success in being applied to computer experiments. Additional references are included for people looking to dive deeper into topics. As a SpringerBrief, we are not attempting to cover all of the topics in this area, but have selected just a few as primary examples of the existing methodology. We hope that this book will be useful for people in a wide variety of fields, including those that study optimization, but also people in fields who use optimization to solve problems in their own disciplines. We have tried to provide sufficient background for statistical, machine learning, and numerical concepts, so that this book can be applicable for people who want to apply these methods to real problems. The book will serve as a useful companion to anyone working in any field that uses computer experiments and computer modeling. Potentially unaware of the topic of Bayesian optimization, practitioners in the field of computer experiments will find Chaps. 3 and 4 of most interest, providing them an introduction to these modern optimization approaches. As computer experiments are increasingly used across a wide range of disciplinary applications, many practitioners have a need for improved optimization methods. Additionally, readers with a background in machine learning, but little to no background in computer experiments, should find the book an interesting case study of the applicability of Bayesian optimization outside of the realm of machine learning.

The book is broken down into four main chapters. Each chapter could stand alone on its own as the topic of an entire book; however, each chapter is built such that the flow of the chapter builds upon the previous one. Chapter 1 introduces the reader to the topic of computer experiments with an abundant variety of examples across many professional industries. Chapter 2 focuses on the task of surrogate model building and contains a mix of several different surrogate models that are used in the computer modeling community, machine learning community, or both. Chapter 3 introduces the core concepts of Bayesian optimization and focuses on the more common case of unconstrained optimization. Chapter 4 finishes the with harder case of constrained optimization and showcases some of the most novel Bayesian optimization methods that currently exist.

Primarily, the book focuses on establishing the essential ideas for both computer experiments and Bayesian optimization such that a reader with limited knowledge of either should be able to pick up the book and start applying its methodology. Stylistically, we prefer to not to distract the reader with code embedded in the body of text in the chapters and instead use the chapters to focus on the foundational ideas and strategies needed to apply the methods. We provide R code as supplementary material for most examples in the book so that the methods mentioned in the book are understandable, reproducible, and transparent. The R code can be found at https://github.com/tpourmohamad/BayesianOptimizationBook.

We are grateful for everyone who has enabled and supported us in this work. The fantastic staff at Springer, especially Editor Laura Briskman, helped bring the idea of this book to fruition. People who provided early stimulating discussions and inspired us to work in this area include Bobby Gramacy, Genetha Gray, and Stefan Wild, with a special thank you to Bobby Gramacy for reviewing a draft of the book and providing many useful comments and suggestions. Finally, we appreciate the support and forbearance of our families as we worked on this book.

South San Francisco, CA, USA   Tony Pourmohamad
Santa Cruz, CA, USA   Herbert K. H. Lee
June 2021

Contents

Chapter 1
Computer Experiments

1.1 Introduction

A computer model (also referred to as computer code or a computer simulator) is a mathematical model that simulates the complex phenomena, or system, under study via a computer program. Controlled experiments, considered to be the gold standard in statistics, are not a viable means for studying complex phenomena when the systems under study are either too expensive, too time consuming, or physical experimentation is simply not possible. For example, weather phenomena, such as hurricanes or global warming, are not reproducible physical experiments, therefore, computer models based on climatology are used to study these events. Another example is the design of a rocket booster and understanding its behavior as it re-enters the atmosphere. It would be prohibitively expensive to build multiple rocket boosters just to study their aerodynamic properties on re-entry, so design is done with computers and re-entry is simulated with a computer model (Gramacy and Lee 2008). Thus, researchers hoping to better understand and model complex phenomena should consider computer modeling as a possible solution.

Computer models have enjoyed a wide range of use, spanning disciplines such as physics, astrophysics, climatology, chemistry, biology, and engineering. At its simplest, a computer model is a mathematical model of the form

$$y = f(x_1, \ldots, x_d) = f(x), \quad x = (x_1, \ldots, x_d)^T \in \mathcal{X}, \tag{1.1}$$

where x is an input variable to the computer model, y is a (possibly multivariate) output from the computer model, and $\mathcal{X}$ is the domain of the input variable. Here, f may or may not have a known analytical representation, and thus the computer model describing the complex system under study may itself also be very complex. Therefore, understanding the computer model can be as challenging of a task as understanding the original physical system it represents. Although possibly stochas-

© The Author(s), under exclusive license to Springer Nature Switzerland AG 2021
T. Pourmohamad, H. K. H. Lee, *Bayesian Optimization with Application to Computer Experiments*, SpringerBriefs in Statistics,
https://doi.org/10.1007/978-3-030-82458-7_1

tic, the focus of this book will be on the case of deterministic computer models where running the computer model for the same input yields the same output always. Other books are available for readers interested in stochastic computer models such as Kleijnen (2015). Much like the design of controlled experiments, one can also construct designed experiments in order to better understand computer models. These experiments, or *computer experiments*, consist of running the computer model at different input configurations in order to build up an understanding of the possible outcomes of the computer model. Here the design of computer experiments can differ substantially from physical experiments and will be elaborated upon in Sect. 1.3.

A common feature of computer models is that they tend to be computationally expensive. Computer models can be extremely complex mathematical programs, and the evaluation of different input configurations may take minutes, hours or even days to calculate a single output. This computational expense makes it prohibitive to try to run the computer experiment at every possible input configuration in order to understand the system. In higher-dimensional input spaces, it will not even be feasible to evaluate the model over a reasonable grid. Thus, a common theme of computer experiments is to try to find an appropriate "cheap-to-compute" model, or *surrogate model*, that resembles the true computer model very closely but is much faster to run. The focus of Chap. 2 is surrogate models, and so we simply mention here that Gaussian processes (Stein 1999) have been used as the typical modeling choice for building statistical surrogate models, and this is due to their flexibility, well-calibrated uncertainty, and analytic properties (Gramacy 2020).

Another typical trait of computer models is that they are often treated as black-box functions. Here, a black-box computer model is a computer model where evaluation requires running computer code that reveals little information about the functional form of the underlying mathematical function, f. The black-box assumption arises due to the fact that f may be extremely complex, analytically intractable, or that access to the internal workings of the computer model are restricted, say, for such reasons as being proprietary software.

One aspect that varies among computer models is whether a function evaluation can be accompanied by derivative (or gradient) information. Some models will provide both an output and derivative information at that point. Other models only provide the output, and cannot provide any derivative information. Classical optimization techniques tend to rely heavily upon derivatives. The focus of this book is on those computer models that do not provide derivative information, which makes optimization more challenging.

Computer models can be built and used to develop methods for computer model emulation (prediction), uncertainty quantification, and calibration, however, the main focus of this book is dealing with computer models for optimization. Computer models are often built with the goal of understanding some physical system of interest, and with that goal usually comes the need to optimize some output of interest from the computer model. For example, in hydrology the minimization of contaminants in rivers and soils is of interest and so computer models representing pump-and-treat remediation plans are often used in order to optimize objectives, such as the associated costs of running pumps for pump-and-treat remediation, while also ensuring that contaminants do not spread (Pourmohamad and Lee 2020). Recalling that most computer models are computationally expensive to run, the need for efficient sequential optimization algorithms (also known as Bayesian optimization (Sect. 3.1)) that do not require many functional evaluations is high, which is why the focus of this book is placed on efficient optimization. More specifically, this book will present solutions to problems of the following form:

$$\min_{x}\{f(x) : x \in \mathcal{X}\} \tag{1.2}$$

or

$$\min_{x}\{f(x) : c(x) \leq 0,\ x \in \mathcal{X}\} \tag{1.3}$$

where $\mathcal{X} \subset \mathbb{R}^d$ is a known, bounded region such that $f : \mathcal{X} \to \mathbb{R}$ denotes a scalar-valued objective function and $c : \mathcal{X} \to \mathbb{R}^m$ denotes a vector of m constraint functions. The unconstrained optimization case in (1.2) will be the subject of Chap. 3, while the constrained optimization in (1.3) will be explored in Chap. 4. Extensions to multivariate objectives are possible. In some cases, the multiple objectives can be collapsed into a joint scalar objective function, such as by taking a linear combination of them. In other cases, the objectives are not comparable, and one is interested in the Pareto frontier of possible points that trade off optimizing the various objectives (Svenson and Santner 2016, for example); these multi-objective problems are beyond the scope of this book.

1.2 Examples of Computer Experiments

The use of computer models spans a vast array of subject areas and each area could potentially be the focus of its own book. In this section, we present a few examples where computer models are used. Our examples are aimed at highlighting the whole gamut of scales at which computer models can be applied, spanning the infinitesimally small to the tremendously large, and a few things in between.

A practical issue with computer models is that there is a dearth of real-world computer models that are freely available and/or accessible to practitioners. This issue arises for a number of reasons, many of which are the same as why most computer models are treated as black-box functions, i.e., the computer model may simply be very complex, analytically intractable, or the software proprietary. When applicable, we try to point out how the examples of computer models can be downloaded and accessed for use.

1.2.1 Groundwater Remediation

Groundwater contamination is one of the most common, and most difficult, environmental problems (Quevauviller 2007). Groundwater is essentially water present below the ground surface, often in underground layers of water-bearing permeable rock, rock fractures, or unconsolidated materials known as aquifers. Groundwater contamination occurs when man-made products such as gasoline, oil, road salts and chemicals get into the groundwater and cause it to become unsafe and unfit for human use. Groundwater remediation is the process that treats polluted groundwater by removing the pollutants and/or converting them into harmless products. The goal of remediation is to make groundwater safe for humans and to minimize the negative impact contaminants have on the environment. To clean polluted groundwater, a variety of physical remediation technologies (e.g., pump-and-treat) have been developed and applied. To better understand how these remediation technologies may work in practice, computer models have been deployed to emulate the physical processes under study. Here we highlight two popular computer models for groundwater remediation.

MODFLOW Originating in the 1980s, the U.S. Geological Survey three-dimensional finite-difference flow model, known as MODFLOW (McDonald and Harbaugh 2003), is a computer model that solves the groundwater flow equation, i.e.,

$$\frac{\partial}{\partial x}\left(K_{xx}\frac{\partial h}{\partial x}\right) + \frac{\partial}{\partial y}\left(K_{yy}\frac{\partial h}{\partial y}\right) + \frac{\partial}{\partial z}\left(K_{zz}\frac{\partial h}{\partial z}\right) + W = S_s\frac{\partial h}{\partial t} \qquad (1.4)$$

where K_{xx}, K_{yy}, and K_{zz} are the values of hydraulic conductivity along the x, y, and z coordinate axes, h is the potentiometric head, W is a volumetric flux per unit volume, S_s is the specific storage of the porous material, and t is time. MODFLOW constructs a finite-difference form of (1.4) in order to solve the partial differential equation. The groundwater flow equation equation describes the flow of groundwater through an aquifer, and so MODFLOW can emulate this groundwater flow process and more. For example, MODFLOW can be used to emulate the flow of groundwater through areas with relatively large hydraulic gradients, such as around the shoreline of a lake, near pumping wells, and along a stream (Fig. 1.1).

MODFLOW is built upon free software that is created and hosted by the U.S. Geological Survey. Many versions of the MODFLOW software exists (MODFLOW 6 is the current core version) and can be downloaded freely at https://www.usgs.gov/software/software-modflow.

MODFLOW can be used to build more complex computer models that depend on the groundwater flow equation. For example, the hydraulic capture problem from the community problems (Mayer et al. 2002) involves a groundwater contamination scenario based on the U.S. Geological Survey computer model MODFLOW. The hydraulic capture computer model is used to help understand how to contain a

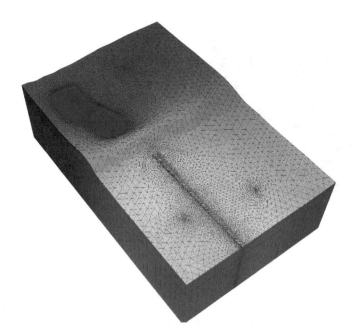

Fig. 1.1 The figure contains a triangular grid in which the size of the triangular cells is reduced in areas with relatively large hydraulic gradients, such as around the shoreline of a lake, near pumping wells, and along a stream. The figure is taken from MODFLOW 6 which is the newest core version

plume of contaminated groundwater by installing up to four wells to control the direction of groundwater movement, and to do so at minimal cost. Analyzed in Lee et al. (2011), Lindberg and Lee (2015), and Pourmohamad and Lee (2016), the hydraulic capture computer model mimics the physical process where the inputs to the computer model are the coordinates of a well and its pumping rate.

Computer modeling is helpful here, because we would not want to run physical experiments to see which configurations contain the contamination and which ones do not, as a failed run would result in further pollution of the area. It is important to do the modeling first via simulation, then only implement a real-world strategy that is likely to be successful in containing the plume of contaminants. While the equations that determine the flow behavior are known and deterministic, the model is treated as a black-box because the equations do not have an analytical solution and so the solution can only be found using numerical approximation. The community problems are an example of constrained optimization, where the objective is to minimize cost subject to the constraint that no pollution escapes the plume. This optimization problem is challenging because the objective function is nonlinear and the constraint function is irregular, with multiple disconnected regions where the constraint is satisfied.

Lockwood Introduced in Matott et al. (2011), the pump-and-treat hydrology problem involves a groundwater contamination scenario based on the Lockwood

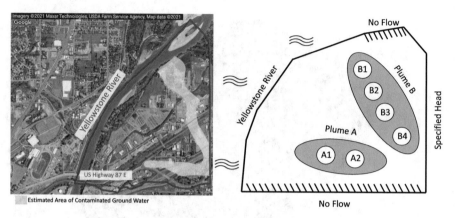

Fig. 1.2 Lockwood site via map (left) and plume diagram (right)

Solvent Groundwater Plume Site located near Billings, Montana. Years of industrial practices in the area have led to the formation of two plumes of chlorinated contaminants, and these two contaminated plumes are slowly, and dangerously, migrating towards the Yellowstone river (Fig. 1.2). To combat the spread of the contaminants, a pump-and-treat remediation plan is proposed, in which pump-and-treat wells are placed at the site of the plumes in order to pump out contaminated water from the soil, purify it, and then return the treated water to the soil. If either of the two plumes reach the Yellowstone river, then the safety of the local water supplies would be compromised, and so preventing both plumes from reaching the Yellowstone river is of critical importance. Thus, the need to deal with the existing problem, and to prevent future problems, led to the creation of the Lockwood computer model.

Like with the MODFLOW problem, the system can be described with deterministic equations, but the solutions are not analytic and require numerical techniques. Computer modeling is critical because a failed experiment would result in contamination of the local water supply.

This computer model was designed to model six pump-and-treat wells at the site of the plumes (two pump-and-treat wells in plume A and four in plume B). The simulator allows modeling of either a simplified problem of fixed well locations and varied pumping rates, or the full problem of variable locations and pumping rates for each of the six wells. The examples in the literature tend to focus on the simpler problem of optimizing pumping rates, which is still a difficult problem. In that case, the goal is to determine the pumping rates at which to set the individual pump-and-treat wells. Thus, the inputs to the computer model are the pumping rates that can be set for the six pump-and-treat wells, i.e., $x = (x_1, x_2, \ldots, x_6)^T \in \mathcal{X}$. Given an input of six pumping rates, x, the output of the computer model is the cost associated with running the pump-and-treat wells, $f(x) \propto \sum_i x_i$, and how much containment escaped each of the plumes, c_1, c_2. Thus, working with the Lockwood computer model can be viewed as an exercise in constrained optimization where the objective of the pump-and-treat hydrology problem is to minimize the cost of running the pump-and-treat wells while containing both plumes. Formally, the objective can be written as:

$$\min_{x}\{f(x) = \sum_{i=1}^{6} x_i : c_1(x) = 0, c_2(x) = 0, \ x \in \mathcal{X}\}. \tag{1.5}$$

While the objective function is actually known in this case, the constraints are non-linear and complex, so computer modeling is needed to understand the constraints.

The code for Lockwood computer model can be downloaded at https://bobby.gramacy.com/surrogates/. See Gramacy (2020) for further details on building the computer model from the code. A number of optimization approaches have been demonstrated on the Lockwood problem (Gramacy et al. 2016; Audet et al. 2018; Pourmohamad and Lee 2020; Gramacy 2020; Pourmohamad and Lee 2021).

1.2.2 Cosmology

Computer models are a powerful and important tool for understanding the cosmos. Computer models play a significant role in cosmology as they provide a means for understanding processes that occur over very long time scales (such as billions of years) and allow for experiments to verify theories, such as, the origin and evolution of the universe (Fig. 1.3).

In order to understand how the universe evolved into what we see today, one needs to follow the non-linear evolution of the cosmos using numerical methods which are typically based on computer simulations. The usual approach, known as the n-body simulation (Fitzpatrick 2012), is to simulate a dynamical system of starting particles usually under the influence of physical forces such as gravity. Thus the process of cosmological simulation starts with some initial conditions according to the dynamics of the particles one wishes to investigate, and then the evolution of these particles is simulated by following the trajectories of each particle under their mutual gravity. The end product is a final particle distribution that can be checked against observations of the universe today. Depending on the number of particles, numerical methods that try to solve the problem directly can become prohibitive or impossible to do without a computer model.

Some of the earliest appearances of computer models in cosmology were for the simulation of star clusters (Aarseth and Hoyle 1963) that used particles in the range of 25–100. Jump ahead roughly 20 years, as computing power and resources increased, cosmological simulations using more than 20,000 starting particles began to appear (Efstathiou and Eastwood 1981). As of today, the Bolshoi simulation (Klypin et al. 2011) is the most accurate cosmological simulation of the evolution of the large-scale structure of the universe. The Bolshoi simulation is a cube roughly 1

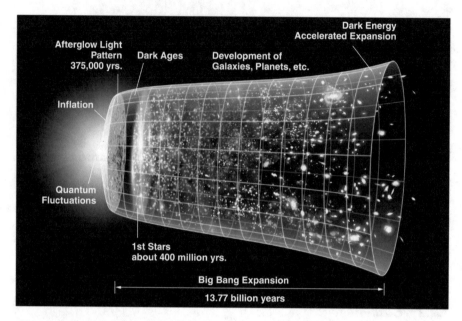

Fig. 1.3 Computer models are used to simulate the evolution of the universe starting with, say, the Big Bang and its expansion

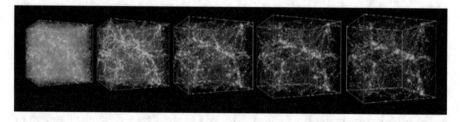

Fig. 1.4 The evolution of the large-scale structures of the universe based on the Bolshoi simulation. Simulations were performed at the National Center for Supercomputer Applications by Andrey Kravtsov (The University of Chicago) and Anatoly Klypin (New Mexico State University)

billion light-years on each side (Fig. 1.4) and is a simulation based on 8,589,934,592 particles, each of which represents about 200 million solar masses.

Increasing the number of particles plays a critical role in helping build better detailed, and often more accurate, simulations of the universe, however, the more particles used the more computer time a simulation consumes. Even with the largest supercomputers nowadays it is only possible to use some million particles if results are to be finished in a reasonable time (say, couple of weeks). The Bolshoi simulation was run on the Pleiades supercomputer, at NASA Ames Research Center, and used 13,824 processor cores, 13 TB of memory, and took 6 million cpu hours to run. Clearly cosmological simulations are extremely computationally expensive and would benefit from a fast surrogate model.

Moreover, most cosmological simulations of the universe depend upon a finite number of cosmological parameters of interest that need to be estimated from an often intractable likelihood function. For example, estimating the parameters of the matter power spectrum (Lawrence et al. 2017) or the 1D flux power spectrum (Rogers et al. 2019). Estimation of these parameters can be carried out via optimization of the intractable (i.e., black-box) likelihood functions. However, this parameter optimization relies upon having data from the extremely computationally expensive cosmological simulations. Thus, obtaining cosmological parameter estimates by optimizing a much cheaper surrogate model of the black-box likelihood function will be essential to reducing the amount of time that the cosmological models take to run.

Computer models for cosmological simulations tend to be abundant, well documented, and freely available for download from the web. A good starting point for learning more about cosmological simulations is the cosmological n-body simulator GADGET which can be downloaded at https://wwwmpa.mpa-garching.mpg.de/gadget/.

1.2.3   Drug Discovery

Discovering new drugs is a complicated, expensive, and time consuming process. Drug discovery is the process by which new candidate medications are discovered. In principle, you can discover or design new drugs by evaluating the fit between a potential drug and its target, via a computer model, in a process called molecular docking (Meng et al. 2011) (Fig. 1.5). A computer model for molecular docking starts out with a well established target protein and binding site, and then tries to fit millions of different chemicals (ligands) into that site, and selects those that work best for further development.

Conceptually, molecular docking can be thought of us as trying to fit two pieces of a jigsaw puzzle together (Fig. 1.5) where, for a given protein puzzle piece, one must investigate hundreds of thousands of ligand puzzle pieces for a potential fit. Thus, molecular docking can be thought of as an optimization problem that describes the "best-fit" orientation of a ligand that binds to a particular protein of interest. However, optimizing molecular docking is a very difficult problem, given the large search space of potential structures coupled with the fact that simulations are almost always computationally expensive. One solution to this problem is to apply Bayesian optimization (the subject of Chap. 3) in order to intelligently search the space of potential chemical structures much faster by adaptively choosing the sequence of molecular compounds to test in drug discovery (Negoescu et al. 2011). Typically this is done by optimizing an objective function that measures how the most probable position, orientation, and conformation with which a ligand can bind to a protein. One example of such an objective function is the Vina scoring function

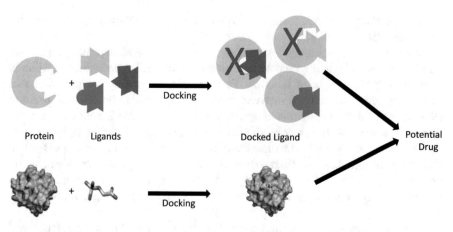

Fig. 1.5 Schematic illustration of docking a small molecule ligand to a protein target producing a stable complex. Potential candidate drugs can be developed from these stable complexes

(Quiroga and Villarreal 2016). The Vina scoring function is an empirical scoring function that calculates the affinity (i.e., a measure of how well a drug binds to the receptor) of protein-ligand binding by summing up the contributions of a number of individual components. The Vina score can be calculated using the Autodock Vina computer model (Trott and Olson 2010), where potential drug candidates with low Vina scores are desirable. Thus, the aim is to minimize the Autodock Vina computer model, where importance is placed on locating particularly low Vina scores efficiently since the input space of potential drug candidates is large.

In general, the main advantage of using computer models for drug discovery relies on the fact that screening of potential candidate drugs can be done much faster on a computer than experimentally in the laboratory. A large number of potential drug targets can be screened with computer models, in a high throughput manner, to quickly identify a handful of effective lead candidates which can then be tested and validated experimentally in a laboratory.

1.2.4 Garden Sprinkler

The garden sprinkler computer model is a multi-objective optimization problem that first appeared in Bebber et al. (2010) and was later adapted to resemble a black-box computer model that can be accessed through the CompModels (Pourmohamad 2020) package in R (R Core Team 2021). The computer model simulates the water consumption associated with using a garden sprinkler, as well as simulating the speed at which the garden sprinkler rotates and the range at which the garden sprinkler can spray water. The aim of the multi-objective optimization problem is to use the garden sprinkler computer model to minimize the sprinkler's water consumption while simultaneously maximizing the speed and range of the sprinkler. There are eight inputs to the computer model based on a set of mathematical equations governing the underlying physics of the computer model. The eight inputs to the computer model are (see Fig. 1.6 as a reference) the vertical (x_1) and tangential (x_2) nozzle angle, the nozzle profile (x_3), the diameter of the sprinkler head (x_4), the dynamic (x_5) and static (x_6) friction moment, the entrance pressure (x_7), and the diameter flow line (x_8). The measurement unit and domain of each computer model input is captured in Table 1.1.

To understand the interplay between the three competing objectives, one could construct the Pareto frontier of all possible points that trade off optimizing the

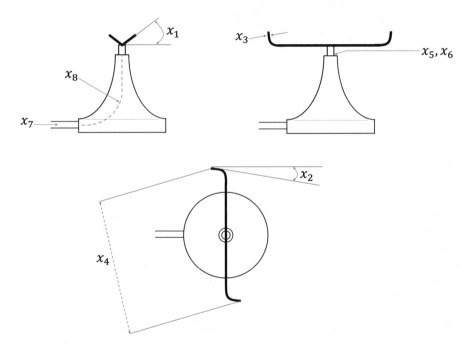

Fig. 1.6 The physical representation of the sprinkler computer model

Table 1.1 Description of the units and domain of the computer model inputs

Attribute	Parameter	Domain	Units
Vertical nozzle angle	x_1	[0, 90]	°
Tangential nozzle angle	x_2	[0, 90]	°
Nozzle profile	x_3	[2e-6, 4e-6]	m²
Diameter of the sprinkler head	x_4	[0.1, 0.2]	mm
Dynamic friction moment	x_5	[0.01, 0.02]	Nm
Static friction moment	x_6	[0.01, 0.02]	Nm/s
Entrance pressure	x_7	[1, 2]	bar
Diameter flow line	x_8	[5, 10]	mm

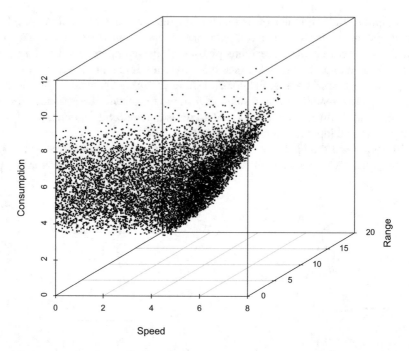

Fig. 1.7 An approximation of the Pareto frontier for the sprinkler computer model

various objectives. Passing a large random sample of the input space to the computer model should help us with approximating the Pareto frontier. Here, for example, we evaluate a rather large random sample, $n = 10,000$, of inputs and obtain the following Pareto frontier approximation (Fig. 1.7). Increases in both the speed and range tends to lead to an increase in water consumption which is in opposition to the goal of the computer model.

Optimizing multi-objective computer models is a difficult task and is outside the scope of this book. Instead, we focus on optimizing single objective functions without regard to the other objective functions values, or subject to the constraint of the other objective function values. For example, one could maximize the range of the garden sprinkler without regards to the water consumption and speed of the garden sprinkler, or one could maximize the range of the garden sprinkler subject to, say, the joint constraints that the water consumption not exceed a certain level and that the speed of the garden sprinkler stay above a certain limit.

Furthermore, optimizing a single objective function (with or without constraints) is a difficult task in its own right. As the dimensionality of the input space increases,

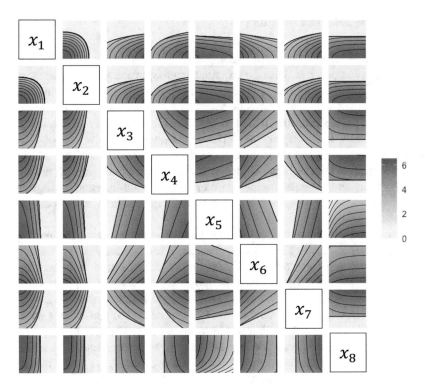

Fig. 1.8 Pairwise contour plots for the objective function corresponding to the range of the garden sprinkler. For each individual pair, two inputs are varied while the other six inputs are held constant at the midpoint of their respective input range

so does the difficulty and complexity of understanding the interplay between inputs and their effect on the objective function output. Exploratory data analysis techniques may not even be feasible for expensive computer models, however, when possible simple plots of the objective function as a function of the inputs may help enhance our understanding of the computer model. For example, in an eight dimensional problem like the garden sprinkler computer model, pairwise contour plots of the objective function as a function of two inputs may yield findings that help us understand which inputs are more influential, or have greater impact, on the computer model output. Figure 1.8 gives an example of pairwise contour plots for the range output of the garden sprinkler computer model. We can see that on average, range tends to increase as nozzle profile (x_3), diameter of head (x_4), and pressure (x_7) increase, and as vertical angle (x_1), tangential angle (x_2), dynamic friction moment (x_5), and static friction moment (x_6) decrease. Diameter of flow line (x_8) appears to have less impact on range, only slightly increasing range as it increases.

1.3   Space-Filling Designs

Computer experiments consist of running the computer model at different input configurations in order to build up an understanding of the possible outcomes of the computer model. Unlike the design and analysis of physical experiments, we are not concerned with issues such as randomization, replication, or blocking (see Joseph (2016) for a discussion) due to the deterministic nature of the computer models. Running the computer model at the same set of inputs always yields the same output, and the order of the runs does not matter. Rather, we should concern ourselves with devising strategies for spreading points (inputs) out evenly throughout the experimental region in order to help build up good surrogates for the computer models. These strategies are commonly referred to as *space-filling designs*.

Intuitively, space-filling designs should result in points that cover the input space well. Likewise, coverage of the input space should become increasingly dense with increasing number of inputs since the domain of the space, X, is bounded. One way to achieve a good space-filling design is to use a Latin hypercube design (LHD) (McKay et al. 1979). Conceptually, random samples taken from a LHD, aptly called Latin hypercube samples (LHS), can be be thought of as a stratified sample of the input space. Latin hypercube samples can be obtained by first dividing the domain of each input variable into n intervals, which we denote $1, \ldots, n$. For each input variable a random permutation of the numbers $1, \ldots, n$ is chosen and the combination of these d permutations form the design. For simplicity, it is often easier to write the LHD with n runs and d input variable as an $n \times d$ matrix, L_D, in which each column is a random permutation of $\{1, \ldots, n\}$. For example, for $d = 2$ and $n = 4$, one potential LHD is $\{1, 4, 3, 2\} \times \{3, 1, 2, 4\}$ or $\{4, 3, 2, 1\} \times \{1, 2, 3, 4\}$ (see Fig. 1.9), i.e., the matrices

$$L_D = \begin{bmatrix} 1 & 3 \\ 4 & 1 \\ 3 & 2 \\ 2 & 4 \end{bmatrix} \quad \text{or} \quad L_D = \begin{bmatrix} 4 & 1 \\ 3 & 2 \\ 2 & 3 \\ 1 & 4 \end{bmatrix}. \tag{1.6}$$

Visually, in two dimensions a LHD is akin to a Latin square in that there will be only one sampled input in each row and column of the grid that is imposed by dividing the domain of each input variable into n intervals (e.g., see Fig. 1.9). In higher dimensions, a LHD becomes a generalisation of this concept to an arbitrary number of dimensions, whereby each sample is the only one in each axis-aligned hyperplane containing it.

Additionally, following the logic and algorithm provided in Fang et al. (2005), if we rescale the domain of each input variable to the unit cube, i.e., define $X = [0, 1]^d$, then we can obtain a LHS by dividing the domain X into n strata of equal marginal probability $1/n$, and sample once from each stratum. This provides us with an easy to follow algorithm for generating a LHS that looks like the following:

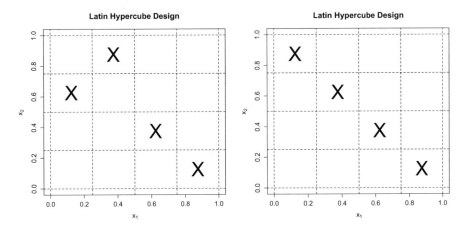

Fig. 1.9 Two different Latin hypercube designs based on $n = 4$ runs and $d = 2$ input variables

Step 1:   Independently take d permutations $l_{i1}, \ldots, l_{in}$ of the integers $1, \ldots, n$ for $i = 1, \ldots, d$.

Step 2:   Take nd uniform variates $u_{ij} \sim U(0, 1)$, $j = 1, \ldots, n$, $i = 1, \ldots, d$, which are mutually independent. Let $x_j = (x_{1j}, \ldots, x_{dj})$, where

$$x_{ij} = \frac{l_{ij} - u_{ij}}{n}.$$

Subtracting random values, $u_{i,j}$, from the uniform distribution ensures that the points will be randomly distributed within their respective interval (e.g., Fig. 1.10). The $n \times d$ matrix $L_S = \{x_1, \ldots, x_n\}$ is a LHS. This two step algorithm is simple to implement in a computer program which is why most statistical software have built-in functions for generating LHS.

In this book, we shall use LHDs as our preferred method for constructing space-filling designs, however, LHDs are by no means the only choice of space-filling designs that exist. Based on the problem at hand, one type of space-filling design may be preferable to another, and there is no shortage of choices. Alternative choices to the LHD include uniform designs (Fang 1980), maximum entropy designs (Shewry and Wynn 1987), integrated mean square error designs (Sacks et al. 1989), minimax and maximin distance designs (Johnson et al. 1990), nested space-filling designs (Qian et al. 2009), sliced space filling designs (Qian and Wu 2009), multi-layer designs (Ba and Joseph 2011), minimum entropy designs (Joseph et al. 2015), bridge designs (Jones et al. 2015), distance based designs (Zhang et al. 2021), and more.

Lastly, you may ask yourself why not just use a simple random sample of the input space? Although a perfectly valid strategy, simple random samples can produce samples that tend to cluster around each other, leading to large gaps in the input space (e.g., Fig. 1.11) and thus suboptimal space-filling designs.

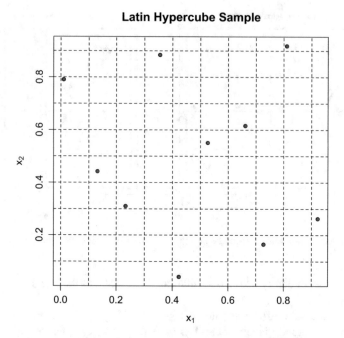

Fig. 1.10 An example of a two dimensional LHS. Note that each row and column, defined by the grid, contains only one input point

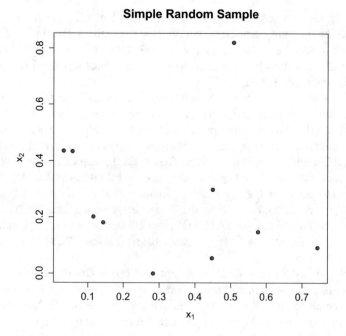

Fig. 1.11 An example of a simple random sample of the input space in two dimensions

However, it is worth noting that even good space-filling designs can lead to poor samples in rare instances. For example, in two dimensions (and generalized to higher dimensions) it is possible to generate a LHS where the points are sampled along the diagonal of the grid (e.g., Fig. 1.9). A design such as this clearly has poor space-filling characteristics which will ultimately impact the quality of the surrogate model.

Chapter 2
Surrogate Models

2.1 Gaussian Processes

Traditionally, the canonical choice for modeling computer experiments has been the Gaussian process (Sacks et al. 1989; Santner et al. 2003; Kleijnen 2015). Gaussian processes were originally developed for geological applications, where a model for spatially correlated data was needed, and one that could either interpolate or smooth the data. The choice of Gaussian processes for deterministic computer simulators was due to this combination of their spatial correlation and their ability to interpolate, as deterministic computer simulators are observed without noise and are generally assumed to have outputs that vary smoothly as a function of their inputs. We will return to both of these assumptions after describing the basic model.

A Gaussian process is a model for a function, in that it is a distribution over functions such that the joint distribution at any finite set of points is a multivariate Gaussian distribution, i.e.,

$$Y_n \sim N(\mu_n, \Sigma_n) \tag{2.1}$$

for points $(x_1, y_1), \ldots, (x_n, y_n)$ where X denotes a location or a set of inputs and Y denotes a response or output. A Gaussian process can be thought of an infinite-dimensional multivariate normal, with values at all points in the X domain. Clearly it is not possible to work directly with an infinite number of points, so a functional form is used instead. The fundamental characterization of a Gaussian process thus requires the specification of a mean function, $\mu(\cdot)$, and a covariance function, $C(\cdot, \cdot)$, which is evaluated at the n points and represented in the covariance matrix Σ_n. Gaussian processes are commonly used in probabilistic modeling when priors over functions are needed and reference to an underlying parametric representation is undesirable. Many references are available for more details on

© The Author(s), under exclusive license to Springer Nature Switzerland AG 2021
T. Pourmohamad, H. K. H. Lee, *Bayesian Optimization with Application
to Computer Experiments*, SpringerBriefs in Statistics,
https://doi.org/10.1007/978-3-030-82458-7_2

Gaussian processes, such as Stein (1999), Rasmussen and Williams (2006), and Gramacy (2020).

In practice, two simplifying assumptions are typically made. The first is stationarity, which is a translation-invariance property: the joint distribution of a set of points is the same after translating each of the points by a fixed amount. For a Gaussian process, this implies that the mean function is constant and the covariance function depends only on the difference vector between two points, not on their specific locations. As many real-world functions have non-constant means, the standard practice is to first subtract off a mean function and then to work with a zero-mean Gaussian process. The second simplifying assumption is isotropy, which means that the covariance between two points depends only on their distance and not on their relative orientation, i.e., the covariance between points is the same in all directions. In the context of computer models, normalizing the inputs so that they are all on the same scale can help make this assumption more reasonable.

Denote the data as $D_n = \{(x_1, y_1), \ldots, (x_n, y_n)\}$. Assuming we have subtracted off any mean function, we work with a zero-mean Gaussian process and we decompose the covariance function into a scale parameter τ and a correlation function $K(\cdot, \cdot)$:

$$Y_n \sim N(0, \tau^2 K_n). \tag{2.2}$$

There are several families of stationary correlation functions that are commonly used, the two most common being the power exponential and the Matérn. The power exponential correlation function, evaluated at two input locations $x_1 = (x_{11}, \ldots, x_{1m})$ and $x_2 = (x_{21}, \ldots, x_{2m})$, looks like

$$K(x_1, x_2) = \exp\left\{ -\sum_{j=1}^{m} (x_{1j} - x_{2j})^p / \theta_j \right\} \tag{2.3}$$

where $0 < p \leq 2$ is the power and $\theta = (\theta_1, \ldots, \theta_m)$ is a vector of scale parameters. In the isotropic case all of the scale parameters are the same, i.e., $\theta_i = \theta_j$ for all i and j. Figure 2.1a shows four realizations in one dimension ($m = 1$) with $p = 2$ and $\theta = 1$. The choice of $p = 2$ is called the squared exponential, or Gaussian, and it produces infinitely differentiable surfaces which are very smooth. This is a common choice for computer experiments. As p decreases, the surfaces become less smooth. Figure 2.1b shows four realizations for $p = 1$ and $\theta = 1$. The power parameter p can be difficult to estimate, so it is usually fixed, and we take $p = 2$ for the rest of this book.

The Matérn correlation function (Stein 1999) for $m = 1$ is

$$K(x_1, x_2) = \frac{2}{\Gamma(v)} \left(\frac{x_1 - x_2}{2\theta} \right)^v B_v \left(\frac{x_1 - x_2}{\theta} \right) \tag{2.4}$$

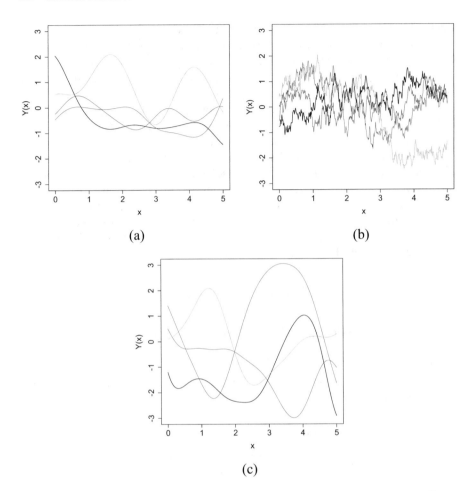

Fig. 2.1 Realizations of Gaussian processes under three different correlation functions. Different correlation functions can produce drastically different processes. (**a**) Squared exponential. (**b**) Powered exponential. (**c**) Matérn

where B_ν is a modified Bessel function of the second kind, ν is a smoothness parameter, and θ is a scale parameter. Figure 2.1c shows four realizations for $\nu = 1$ and $\theta = 1$. ν is often fixed in this context, as it can be difficult to estimate. This correlation function is generally preferred in the spatial statistics literature (Banerjee et al. 2004).

When fit to data, the above formulation will interpolate, that is, it treats the data as if there is no noise, which makes some intuitive sense for deterministic computer experiments where running the code twice at the same inputs will always produce the same output. However, there can be numerical stability issues, strong dependence on assumptions being exactly correct, and a need to ensure adequate sampling of the space in order for interpolation to work reliably. In practice, it is usually better to include a small amount of noise in the model, which is done via a nugget (Gramacy and Lee 2012), which adds a small value, η, to the diagonal of the correlation matrix K. While this means K is no longer a proper correlation matrix (because the diagonal is larger than 1), we can still use it the same way in all of the computations, and we will make this slight abuse of notation and continue to refer to it as a correlation matrix. There is a reparametrization of the nugget model that is equivalent to a model that observes Y with noise, but the nugget parameterization is more common in the computer model literature.

Throughout this book, we work within the paradigm of Bayesian statistics, where prior beliefs are updated with the data to get a posterior distribution representing the full uncertainty of the estimates. Readers who need more background on Bayesian statistics may find texts such as Hoff (2009) and Gill (2014) helpful. Thus to fit the parameters of a Gaussian process, we first put priors on the parameters τ^2 and $\psi = (\theta, \eta)$, $p(\tau^2)$ and $p(\psi)$. We then update the priors with the likelihood in Eq. 2.2, using Bayes' Theorem, to get the posterior

$$p(\tau^2, \theta | Y_n) \propto |\tau^2 K_n|^{-\frac{1}{2}} \exp\left(-\frac{1}{2\tau^2} Y_n^T K_n^{-1} Y_n\right) \times p(\tau^2) p(\psi). \qquad (2.5)$$

A typical choice of prior for τ^2 is the reference prior $p(\tau^2) \propto 1/\tau^2$ (Berger et al. 2000).

The marginal predictive distribution, $p(Y(x)|D_n, \tau^2, \psi)$, for a new observation at input x is a Student's t-distribution with n degrees of freedom, and mean, $\mu(x)$, and scale, $\sigma^2(x)$, given by

$$\mu(x) = k_n^T(x) K_n^{-1} Y_n \qquad (2.6)$$

$$\sigma^2(x) = \frac{Y_n^T K_n^{-1} Y_n (K(x, x) - k_n^T(x) K_n^{-1} k_n(x))}{n}. \qquad (2.7)$$

As an example, Fig. 2.2 shows the result of fitting a Gaussian process with a squared exponential correlation function to seven points sampled from the function

$$f(x) = \exp(-1.4x) \cos(7\pi x/2).$$

For this example, the length scale parameter is $\theta = 0.05$ and a nugget of 1e-6 is used. The dark line shows the true function f and the dashed line shows the fitted mean from Eq. (2.7), although it is difficult to distinguish the two lines because they are nearly identical for most of the plotted area. The lighter gray lines are 100 draws from the predictive distribution, showing that the predictive variability increases away from observed data locations and shrinks to the nugget at the data locations. On the left side, where it is predicting beyond the range of the data, the variability rapidly increases outside the range of the data.

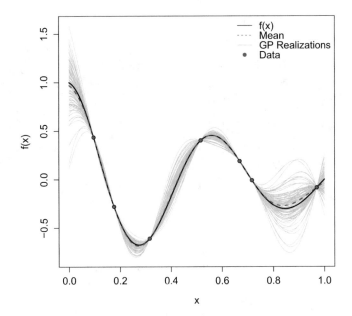

Fig. 2.2 Gaussian process fit and 100 draws from the predictive distribution

A two-dimensional example is given in Fig. 2.3 for the function

$$f(x_1, x_2) = -(\cos((x_1 - 0.1)x_2))^2 - x_1 \sin(3x_1 + x_2).$$

Because the sampled locations do not fully cover the region, the fitted mean function is close to the truth but shows some discrepancies that would diminish as more locations are sampled. The predictive variability is again larger for locations farther from sampled locations.

In the context of optimization, we are not necessarily focused on marginal predictions, because we are typically using the Gaussian process inside of a subroutine. In that setting, we more frequently are interested in posterior distributions and posterior predictive distributions conditional on the values of some of the parameters. We may fit the values of τ^2 and $\psi = (\theta, \eta)$ within the subroutine, and then be interested in the predictive mean and predictive variance conditional on τ^2 and ψ. In that case, the conditional predictive distribution is a Gaussian distribution. Figure 2.4 shows 100 draws from this conditional predictive distribution. Compared to Fig. 2.2, the conditional distribution has a little less variability, which can be seen in the decreased spread of the gray lines.

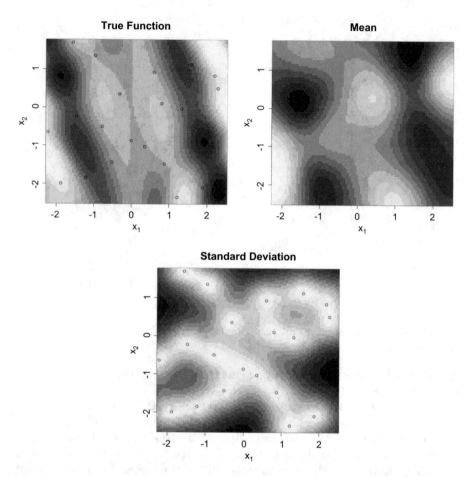

Fig. 2.3 2-d example of a Gaussian process fit to the true function in the left plot. The middle and right plots show the fitted mean and predictive standard deviation functions

2.2 Treed Gaussian Processes

Gaussian processes have become popular in computer modeling because they are both fairly flexible and fairly tractable, and they have worked well in a wide variety of applications. However, there are some simulators where the modeling assumption of stationarity is not a good assumption, and so the standard Gaussian process may not work as well. While there is an increasing literature on fully nonstationary Gaussian processes, most computer simulators don't require complete nonstationarity, often just a little bit will suffice. This may be because of a small number of discontinuities, or differing behavior in a couple of important regions of the space. A relatively straightforward extension of a Gaussian process is a treed Gaussian process (Gramacy and Lee 2008). The basic idea is to partition the input

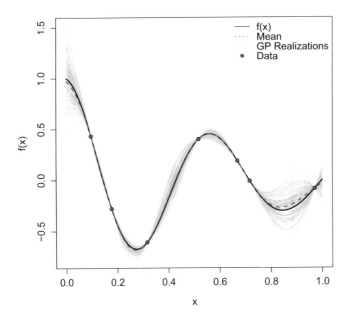

Fig. 2.4 Gaussian process fit and 100 draws from the conditional predictive distribution

space into a relatively small number of regions and to fit independent stationary Gaussian processes in each region. When done within the Bayesian framework, the partitions can be inferred, and the resulting predictive function can be averaged across the posterior for the partitioning, allowing for a smooth posterior predictive function when the truth is smooth, and allowing for a discontinuous predictive function when the truth is discontinuous.

Partitioning is based on the approach from the tree models of Chipman et al. (1998) and Chipman et al. (2002), which defines a recursive binary tree-generating prior, where each node has an independent probability of splitting into two partitions, with the splitting probability decreasing with the depth of the node. The default prior splitting probability is $(1 + q)^{-2}/2$, where q is the depth of the node. This prior tends to produce relatively small trees, which are all that are needed when combined with the flexibility of Gaussian processes. Posterior inference is done with Markov chain Monte Carlo, inferring both the tree structure and the component Gaussian processes simultaneously. Implementation details are available in Gramacy and Lee (2008), and a good implementation is the R package `tgp` (Gramacy 2007; Gramacy and Taddy 2010).

Fitting a treed Gaussian process to the data in Fig. 2.4 should lead to essentially the same fit as a standard Gaussian process, because a stationary model works well on this dataset, so the treed model doesn't partition and just fits a single Gaussian process to all of the data. Using the `tgp` R package, and fixing the nugget to the same 1e-6 as before, and using a prior that puts the length scale parameter close to the previous 0.05, Fig. 2.5 shows the treed Gaussian process fit and pointwise

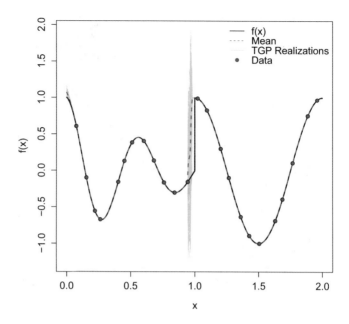

Fig. 2.7 Treed Gaussian process fit and 100 pointwise draws from the predictive distribution for a discontinuous true function

2.3 Radial Basis Functions

Although Gaussian processes have become the canonical choice for modeling computer experiments, they are by no means the only modeling choice available. Here we describe one more potential surrogate model that has substantial roots in the machine learning literature (see Chapter 5 of Bishop (1995)). Given the deterministic nature of computer models, another potential choice of exact interpolating models is the radial basis function (RBF) regression model. Tracing its roots back to early works such as Hardy (1971), radial basis function regression creates an interpolating model through a linear combination of radial functions. A radial basis function is defined as any real-valued function $\varphi : \mathbb{R}^d \to \mathbb{R}$ for which its function value only depends on the magnitude of its argument, i.e.,

$$\varphi(x) = \varphi(||x - a||) \tag{2.8}$$

where a is some fixed point called a center. Setting $a = 0$ produces a function that is radial around its origin. Moreover, any function $\varphi(x)$ that satisfies the property $\varphi(x) = \varphi(||x||)$ is called a radial function. Typically, $||x||$ is taken to be Euclidean distance, however, other metrics may be used.

Now, in the context of computer experiments, recall that we have data $D_n = \{(x_1, y_1), \ldots, (x_n, y_n)\}$, where $y_i = f(x_i)$ is the i-th output of running the computer model at the i-th input for $i = 1, \ldots, n$. Given the deterministic nature of the computer model, our goal is to find an interpolating function $s(x)$ such that $s(x_i) = f(x_i)$. The radial basis function regression model constructs the linear interpolation function, $s(x)$, through a linear combination of radial basis functions, i.e.,

$$s(x) = \sum_{i=1}^{n} \lambda_i \varphi(||x - x_i||), \qquad (2.9)$$

where $\lambda_i \in \mathbb{R}$ is a weight and $x \in \mathbb{R}^d$. Given the condition that the model must interpolate at observed data, i.e., $s(x_i) = f(x_i)$, we can rewrite (2.9) in matrix notation as

$$\underbrace{\begin{bmatrix} \varphi(||x_1 - x_1||) & \varphi(||x_1 - x_2||) & \cdots & \varphi(||x_n - x_1||) \\ \varphi(||x_1 - x_2||) & \varphi(||x_2 - x_2||) & \cdots & \varphi(||x_n - x_2||) \\ \vdots & \vdots & \ddots & \vdots \\ \varphi(||x_1 - x_n||) & \varphi(||x_1 - x_n||) & \cdots & \varphi(||x_n - x_n||) \end{bmatrix}}_{\Phi_{n \times n}} \underbrace{\begin{bmatrix} \lambda_1 \\ \lambda_2 \\ \vdots \\ \lambda_n \end{bmatrix}}_{\Lambda_{n \times 1}} = \underbrace{\begin{bmatrix} f(x_1) \\ f(x_2) \\ \vdots \\ f(x_n) \end{bmatrix}}_{F_{n \times 1}}, \qquad (2.10)$$

or more compactly in matrix form as

$$\Phi \Lambda = F. \qquad (2.11)$$

It is obvious that the $n \times n$ matrix Φ is symmetric, and that for (2.11) to have a unique solution, we require that Φ be non-singular so that $\Lambda = \Phi^{-1} F$. Solving for Λ is straightforward, however, there are no explicit instructions for specifying the form of the radial basis function φ. Subject matter expertise and prior knowledge may be used to help inform the type of radial basis function specified based on knowledge of, say, smoothness properties, differentiability, or decay of the function f. Table 2.1 lists some of the commonly used types of radial basis functions.

Most radial basis functions contain parameters, e.g., c, that can be used to adjust the shape of the radial basis functions. For example, Fig. 2.8 demonstrates a few of the common radial basis function choices for varying values of parameter c.

Note that the Gaussian radial basis function is related to the power exponential correlation function (Sect. 2.1) when the power $p = 2$, and thus shares the same properties such as infinitely differentiable surfaces. Furthermore, although not discussed in this book, radial basis functions share a rich history with Gaussian processes as well as they can be used to specify the kernel of the convolution in process convolution Gaussian process models (Higdon 1998, 2002).

Revisiting the one dimensional example in Sects. 2.1 and 2.2, consider the function

$$f(x) = \exp(-1.4x) \cos(7\pi x/2)$$

Table 2.1   Some common radial basis functions

Radial basis function	$\varphi(r)$	Parameter
Gaussian	$e^{-(cr)^2}$	$c > 0$
Multiquadric	$\sqrt{r^2 + c^2}$	$c > 0$
Inverse quadratic	$\frac{1}{r^2+c^2}$	$c > 0$
Inverse multiquadric	$\frac{1}{\sqrt{r^2+c^2}}$	$c > 0$
Polyharmonic spline	r^c	$c = 1, 3, 5, \ldots$
	$r^c \ln(r)$	$c = 2, 4, 6, \ldots$
Thin plate spline	$r^2 \ln(r)$	–
Bump function	$\begin{cases} \exp\left(\frac{1}{1-(cr)^2}\right) & \text{if } r < \frac{1}{c} \\ 0 & \text{otherwise} \end{cases}$	$c > 0$

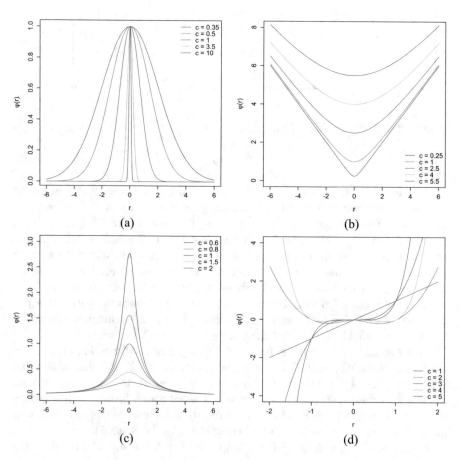

Fig. 2.8   Examples of different radial basis functions and the effect of varying the shape parameter c. (**a**) Gaussian basis functions. (**b**) Multiquadric basis functions. (**c**) Inverse multiquadric basis functions. (**d**) Polyharmonic splines

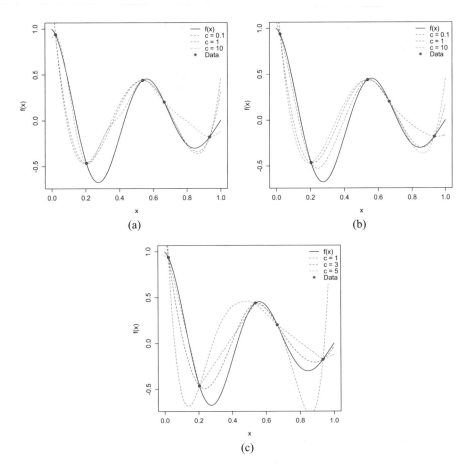

Fig. 2.9 Examples of prediction from a radial basis function regression model under three different radial basis function specifications with varying values of parameter c. (**a**) Gaussian RBF. (**b**) Multiquadratic RBF. (**c**) Polyharmonic splines RBF

over the input space $\mathcal{X} \in [0, 1]$. In this example, we evaluate the function f at a smaller number of inputs (here we have chosen five points from a LHS) in order to highlight the effect of varying the parameter c on the interpolation of the radial basis function regression model. With the exception of the polyharmonic spline RBF with $c = 1$ (i.e., a piecewise linear interpolator), all of the following examples of interpolation under the chosen RBFs would have looked near identical for as few as seven or more data points. Now, for sake of example, we show the surrogate model prediction under the Gaussian, multiquadratic, and polyharmonic spline RBF for varying values of c (Fig. 2.9).

As we should expect, all predicted (dashed) lines interpolate exactly at observed data points, and for different values of c there is a wide range of behaviors between surrogate models. What we tend to see visually is that the parameter c plays a similar role to the length scale parameter in a Gaussian process correlation function in that it dictates "how close" two data points have to be to influence each other significantly.

Chapter 3
Unconstrained Optimization

3.1 Bayesian Optimization

A method that dates back to Mockus et al. (1978), Bayesian optimization (BO) is a sequential design strategy for efficiently optimizing black-box functions, in few steps, that does not require gradient information (Brochu et al. 2010). More specifically, BO seeks to solve optimization problems of the form

$$x^* = \underset{x \in \mathcal{X}}{\operatorname{argmin}} f(x). \tag{3.1}$$

The minimization problem in (3.1) is solved by iteratively developing a statistical surrogate model of the unknown objective function f, and at each step of this iterative process, using predictions from the statistical surrogate model to maximize an acquisition (or utility) function, $a(x)$, that measures how promising each location in the input space, $x \in \mathcal{X} \subset \mathbb{R}^d$, is if it were to be the next chosen point to evaluate. Thus, the role of the acquisition function, $a(x)$, is to guide the search for the solution to (3.1). We defer further discussions of acquisition functions to Sect. 3.2, but conceptually it is clear that different choices of acquisition functions should lead to different measures of belief of the search algorithm when searching for the best next input to evaluate. Gaussian processes have been the typical choice of statistical surrogate model for the objective function f in BO, and this is due to their flexibility, well-calibrated uncertainty, and analytic properties (Gramacy 2020).

Practically, what does this all mean? Recalling the idea of groundwater remediation (Sect. 1.2), let us consider a simple toy example where the objective is to drill down into the earth to locate the largest plume of chemical contaminants possible. Why would this be important to do? Well, if we can hypothetically only place one pump-and-treat well in the area of interest then we may want to place it at the site of the area with the largest amount of contaminants in the earth. In practice we would not know the true distribution of the contaminants but, for sake

© The Author(s), under exclusive license to Springer Nature Switzerland AG 2021
T. Pourmohamad, H. K. H. Lee, *Bayesian Optimization with Application to Computer Experiments*, SpringerBriefs in Statistics,
https://doi.org/10.1007/978-3-030-82458-7_3

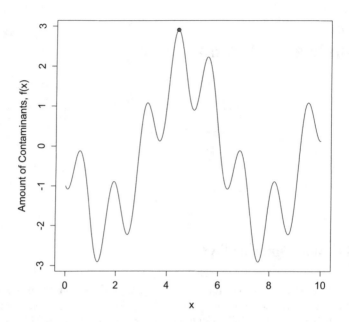

Fig. 3.1 The objective function $f(x)$ to maximize for the ground contaminant example. Here, the global maximum (the red point) is obtained at $x = 4.42124$

of example, let us suppose that the distribution of log concentration of contaminants $f(x)$ looks something like the following (Fig. 3.1). For now, let us not worry about the interpretation of the units on the x and y axes. Although relatively smooth, the function has multiple local maxima, and sharp variation in the output axis, which makes the function hard to optimize.

Initially, we have no idea about what the distribution of the contaminants looks like. We can learn about the distribution of the contaminants by drilling in the earth at different locations. However, this drilling will be costly. Thus, we want to minimize the number of drillings required while still finding the location of the maximum amount of soil contaminants. Now, drilling can continue to be done over the entire domain, $\mathcal{X}$, so that we can we can learn the true function, f, very well, but that seems rather wasteful given that drilling is expensive. Furthermore, we are primarily interested in locating the maximum amount of contaminants, and so we are not actually interested in learning what the true function looks like in areas where contaminants are low and so we should not spend a significant amount of resources in those areas if we do not need to. Instead, based on our current knowledge of the true function, we should try to balance investigating and drilling in uncertain regions which might unexpectedly have high levels of contaminants (exploration), versus focusing on regions we already know have higher contaminant levels (exploitation). This is the fundamental concept of Bayesian optimization; how to choose the next point to evaluate in order to balance exploration—improving the surrogate model in the less explored parts of the search space, and exploitation—favoring parts the surrogate model predicts as promising. Again, we defer this discussion to Sect. 3.2, but it will be shown that maximization of a well-chosen acquisition function can be used within BO to guide the search. Lastly, we summarize here the general characteristics of BO and relate the drilling example to those characteristics:

General characteristics of BO	Characteristics of the drilling example
The domain of f is simple	The domain is a simple box constraint, i.e., $0 \le x \le 10$
f is continuous but lacks special structure	f is neither a convex nor concave function which results in many local maxima
f is expensive to evaluate	Drilling into the earth is costly
Evaluations of f are derivative free	Drilling into the earth provides no gradient information

Now there is no reason why BO could not be applied to the actual physical experiment of drilling and searching for the contaminants, however, the process is expensive and so running the physical experiment would be very costly. Instead, it would be useful to construct a computer model of the physical process based on the physics of the environmental system under study and subject matter expertise. BO could then be used to optimize the output of the computer model which would likely result in savings of both money and time for those who are doing the drilling. Once the best solution to the BO problem has been found, the solution can be verified by running the actual physical experiment of drilling. At this point, you may be asking yourself what happens if the solution from the BO does not match the reality of the physical experiment? Well that is not a problem with the BO algorithm, but rather a problem with the accuracy of the computer model at matching reality. Remember that computer models are constructed and used when physical experimentation is not feasible due to a variety of constraints (Sect. 1.1), and so our reliance on the computer model is to treat it as the true state of reality when optimizing it. Of course we should not ignore the discrepancies between the computer model and reality, and again this is not a problem of optimization but rather a problem in what is known as computer model calibration (Kennedy and O'Hagan 2001). In simplest terms, computer model calibration is a technique that models a discrepancy term between the output of the computer model and reality, which can then be used to make better (i.e., calibrated) predictions of reality. Although computer model calibration is an important topic (entire chapters have been devoted to it, e.g., Santner et al. (2003, Chapter 8) and Gramacy (2020, Chapter 8)), it is outside the scope of this book and so we assume that the computer models that we deal with in this book are either well calibrated or are the only means of experimentation that we have access to.

Describing BO as a simple algorithm, we present the following general formulation for solving problems of the form (3.1):

Algorithm 1: The general BO algorithm

Initialization:

  Start with an initial data set D_0

  for $n = 1, \ldots, N$ **do**

  | Fit a Gaussian process surrogate model to D_{n-1}, i.e., $Y(x)|D_{n-1} \sim$ GP;
  | Select $x_n = \text{argmax}_{x \in \mathcal{X}} a_{n-1}(x)$;
  | Evaluate f at x_n to obtain y_n;
  | Augment data $D_n = D_{n-1} \cup \{(x_n, y_n)\}$;

  end

  Return:

    $x^* = \text{argmin}_{x \in \mathcal{X}} f(x)$

Following the algorithm, we initialize the BO algorithm by collecting a set of outputs evaluated at a small set of inputs. Here, small is defined by the problem at hand and the computational budget. As the number of input dimensions increases, it is common to also increase the starting sample size. Recalling Sect. 1.3, we recommend collecting an initial data set based on a space-filling design. Our default space-filling design recommendation for BO is the Latin hypercube design. After initialization, the algorithm proceeds in an iterative fashion selecting new inputs to evaluate based on maximizing a prespecified acquisition function that encapsulates our beliefs in how the input space should be searched. A key step in the algorithm, a Gaussian process surrogate model is constructed to model the objective function f, and the predictions that come out of this Gaussian process surrogate model are then used in the maximization of the acquisition function. Based on the acquisition function, we then evaluate the objective function at the best next point and update our current data to include this evaluation. The procedure continues based upon computational budget (such as time, resources, money, etc.) and terminates once all such budget is exhausted.

At this point, you may be wondering where is the Bayesian in Bayesian optimization? While not all BO algorithms are actually Bayesian, the term comes from the Bayesian updating of the Gaussian process in the original formulation, which is what we are using here. At each iteration, the computer model is run and a new input-output pair is observed. The Gaussian process surrogate model is then updated with the new observation, leading to a new posterior distribution and posterior predictive distribution. This posterior predictive distribution is then used in the acquisition functions that follow.

Now of course, BO is not the only way to solve optimization problems like those in (3.1). Conventional local optimization methods like the Nelder-Mead method (Nelder and Mead 1965) or the Broyden–Fletcher–Goldfarb–Shanno (BFGS) algorithm (Broyden 1970; Fletcher 1970; Goldfarb 1970; Shanno 1970), and the alike, can be used to solve optimization problems of the same form. However, local optimization methods suffer from the distinct problem that they tend to get trapped in local modes of the objective function. Local optimization methods typically only guarantee local solutions rather than global ones, and so being able to escape local modes of the objective function is of little importance to the search algorithm. On the other hand, BO algorithms are capable of taking both a local and global perspective to the optimization problem, and so this issue of getting trapped in local modes is less of a problem for BO.

3.2 The Role of the Acquisition Function

At the heart of all Bayesian optimization algorithms is an acquisition function, $a(x)$, for effectively guiding the search. The acquisition function serves as a tool for encapsulating our beliefs about how to best choose the potential candidate inputs to evaluate the objective function at. Often regarded as an "inner" optimization problem since it is embedded within the optimization problem in (3.1), we can select candidate inputs for minimizing $f(x)$ by finding the input that maximizes the acquisition function, i.e.,

$$x_n = \operatorname*{argmax}_{x \in X} a_{n-1}(x). \tag{3.2}$$

Typically, the acquisition function is thought of as a utility function and so finding the input that maximizes the acquisition function can be regarded as evaluating the utility of one input over another for minimizing $f(x)$. For example, one could consider an acquisition function based solely on the predictive mean of the Gaussian process surrogate model, i.e.,

$$a(x) = -\mu_n(x). \tag{3.3}$$

Choosing the input that maximizes this acquisition function is the same as picking the lowest point on the mean predictive surface under the Gaussian process. Thus, an acquisition function such as this would place high utility on inputs where the Gaussian process model predicts as promising. Although this may work in instances when the Gaussian process model does well at predicting the objective function across the entire input space, it turns out that this is a poor strategy in general as the acquisition will tend to get stuck searching in areas it thinks are promising rather than exploring the input space globally. Acquisition functions such as these are referred to as *greedy* algorithms and suffer from the poor behavior of getting trapped in local optima. It turns out that a better solution is to use acquisition functions that balance exploration (improving the model in the less explored parts of the search space) and exploitation (favoring parts the model predicts as promising). This exploration-exploitation trade-off helps to ensure that the acquisition function values candidate points both locally and globally.

Now, different acquisition functions can lead to different ways of assigning utility to candidate inputs. And unfortunately, there is no one size fits all acquisition function for every type of objective function. However, there are several good choices of acquisition functions that exist in the literature, of which some of we shall expand upon in Sect. 3.3.

3.3 Choice of Acquisition Function

A large body of active research in the BO community is devoted to the development of novel acquisition functions. In this section we present some of the most commonly used acquisition functions in the BO literature. Each acquisition function will rely on the fact that the underlying black-box function, $f(x)$, is being modeled by a Gaussian process $Y(x)$. Recall that although the black-box functions we seek to optimize are deterministic, it is typically advantageous in practice to use a small nugget term, η (see Sect. 2.1), when modeling the black-box functions (Gramacy and Lee 2012). This noise term η will lead to models that do not exactly interpolate at observed inputs, and so a large nugget term does not make much sense to use when the black-box functions are deterministic. Practically speaking, there are two ways to ensure that a small nugget term is used. One is to place a prior on the nugget parameter η that restricts the nugget to small values, or the other option is to simply fix the nugget at a reasonably small value based on the scale of the data. In the implementation of the analyses that follow, we take the latter approach.

Lastly, while not explored in this book, the following acquisition functions presented in the chapter will need modifications when the objective function is a stochastic black-box function. Readers interested in BO of stochastic black-box functions are encouraged to see Picheny et al. (2013), Picheny and Ginsbourger (2014), Jalali et al. (2017), and Baker et al. (2021), for example.

3.3.1 Probability of Improvement

Perhaps the simplest, and most intuitive, of all the the acquisition functions is the probability of improvement (PI) acquisition function (Kushner 1964). Conceptually, as its name suggests, the PI acquisition function chooses the best next point to evaluate by picking the point that has the highest probability of improving upon the current best solution. Here, the current best solution, at the nth iteration of the search, is defined to be $f^n_{\min} = \min\{f(x_1), \ldots, f(x_n)\}$, i.e., the minimum value of all the evaluated outputs. In order to make probabilistic statements about some output $f(x)$ being lower than $f_{\min}$ we need to assume a distribution for the underlying function f. Recall that we can model the black-box function, $f(x)$, by a Gaussian process $Y(x)$ where, conditional on all of the data D_n evaluated at the nth iteration, the predictive distribution of $Y(x)|D_n$ is a normal distribution with with mean and standard deviation, $\mu_n(x)$ and $\sigma_n(x)$, respectively. Thus, the PI acquisition function can be defined as

$$a_{PI}(x) = \Pr(Y(x) \leq f_{\min}^n)$$

$$= \int_{-\infty}^{f_{\min}^n} N(Y(x); \mu_n(x), \sigma_n(x))dY(x)$$

$$= \Phi\left(\frac{f_{\min}^n - \mu_n(x)}{\sigma_n(x)}\right) \tag{3.4}$$

where $\Phi(\cdot)$ is the standard normal cumulative distribution function (cdf).

Optimization of the objective function can proceed sequentially by maximizing $a_{PI}(x)$ and using the argument that maximizes that function as the next input to evaluate. Perhaps not obvious yet, it turns out that $a_{PI}(x)$ tends to be a "greedy" acquisition function in that it favors exploring the input space where variability is low and promising solutions are high. Although quick to converge to a solution of the optimization problem, this greedy nature can lead to the unfavorable quality of getting stuck in local optima.

Example To better understand idea and implementation of the PI acquisition function, consider the following example of a simple one dimensional objective function from Chap. 2:

$$f(x) = \exp(-1.4x)\cos(7\pi x/2),$$

where $x \in [0, 1]$. Of course in practice we would likely not know the functional form of our computer model f, as well the function would be expensive to evaluate, but for sake of example we do. Now imagine that we have completed the sixth iteration of the BO algorithm (i.e., we have run the computer model 6 times) and want to decide where to choose the seventh input. The left panel in Fig. 3.2 shows the current predictive surface in terms of mean and approximate 95% predictive intervals. The predictive mean surface contains two minima, roughly at $x = 0.25$ and $x = 0.78$, with one of the two minima predicted to be much lower than the other. The right panel in Fig. 3.2 recapitulates the message taken from the predictive mean where we see there are two bumps at $x = 0.25$ and $x = 0.78$. The PI acquisition places much higher probability at $x = 0.25$ as should be expected given the predictive surface. Thus, at the seventh iteration of the algorithm the point at $x = 0.25$ would be selected as being the best next input to evaluate.

Perhaps computational budget still remains and we have decided to choose an eighth input to evaluate. Having just evaluated $x = 0.25$ in the last round of iterations, we see that the predictive surface and probability of improvement changes quite a bit (Fig. 3.3). Given the location of the seventh data point, the uncertainty in the region has shrunk substantially, while the uncertainty around the other local minimum remains high. The greedy nature of the PI acquisition function makes

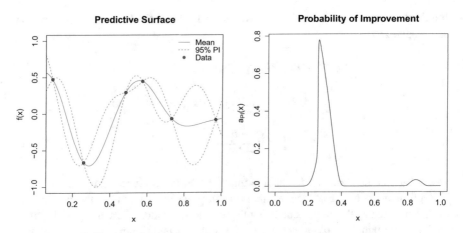

Fig. 3.2 An illustration of the $a_{PI}(x)$ acquisition function at work. The corresponding predictive surface (left) and PI surface (right) given 6 data points

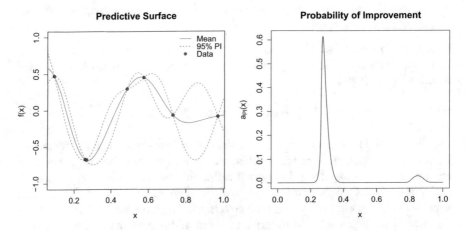

Fig. 3.3 The corresponding predictive surface (left) and PI surface (right) given 7 data points

the BO algorithm favor exploiting the fact that it is very certain it is near the global solution to the problem which is why the probability of improvement further concentrates in the region near $x = 0.25$. Continuing to run the BO algorithm until its budget is exhausted would continue to yield evaluations in the already exploited region.

Example To give an example from start to finish, let us return to the drilling for contaminants problem of Sect. 3.1. Here, we wish to maximize the objective function since we wish to locate the area of maximum contaminants, and so to be consistent with our formulation of our problems as minimizations, we shall instead minimize the negative objective function in order to find the input that maximizes it. To initialize the BO algorithm, the size of the initial set of inputs must first be determined. In practice, there is no rule for selecting the size of this initial data set, but rather it should be based on the complexity of the objective function, as well as the dimension of the input space. Objective functions that are smooth, well behaved, or with few local minima (i.e., low complexity) typically require fewer starting inputs than a more complex function. Likewise, with increasing dimensionality of the input space comes the need to have a larger set of starting inputs. However, given the black-box nature of the objective function, it may not always be feasible to know exactly how complex the underlying problem is. Subject matter expertise may need to be elicited from those in the field in which the computer model is developed in, however, sometimes a best guess at a reasonable starting sample size is the best that can be done.

For now, let us consider the case where we have little to no available information about the complexity of the underlying function for contamination. Given the low dimensionality of the input space (i.e., one dimension), we select an initial set of $n = 10$ inputs based on a LHD. Next, we fit our Gaussian process model to the ten input-output data pairs in order to obtain estimates of the mean and variances of the predictive surface over the entire input space (Fig. 3.4). These estimates can then be plugged into (3.4) in order to select the input that maximizes the PI acquisition function. The true underlying objective function is in black in Fig. 3.4 while the predictive surface is in red. What we can see is that with only ten points, our Gaussian process model is not able to capture the wiggliness of the objective function well which results in our PI acquisition function placing high probability of improvement at one of the local minima.

Given the greedy nature of the PI acquisition function, the BO algorithm will tend to be limited in its exploration of the input space. After initialization, the PI acquisition function will tend to get trapped in the area of whichever minima (local

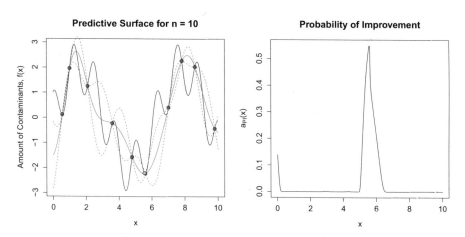

Fig. 3.4 The corresponding predictive surface (left) and PI surface (right) given 10 data points for the drilling example

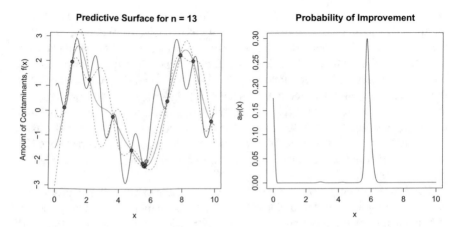

Fig. 3.5 The corresponding predictive surface (left) and PI surface (right) given 13 data points for the drilling example

or global) has a data point close to it, and is the lowest objective value observed. Here in Fig. 3.4, the blue point corresponds to the output of evaluating the input that maximizes the PI acquisition function. The point is within the valley of a local minima, and so unfortunately the BO algorithm will get stuck greedily choosing points around that local minima. As we can see in Fig. 3.5, the next three inputs selected by the PI acquisition function are all within the area of the same local minimum. The probability of improvement over the entire input space does indeed change, but not enough to have a meaningful impact on the escaping the local minimum.

Moreover, something even worse can occur within the BO algorithm. As we accumulate more and more data within the area of the local minimum, the estimation of the parameters of the correlation matrix of the Gaussian process can also suffer. For example, after initialization of the BO algorithm the next five inputs selected are all very close to one in another, being positioned around a local minimum (Fig. 3.6). As more and more inputs are evaluated within close proximity of each other, what happens in this particular example is that the estimation of the length scale parameter is inadequate for capturing the global shape of the acquisition function, which in turn leads to a very poor predictive surface and poor uncertainty quantification. The fitted length scale in Fig. 3.6 has become small, resulting in a predicted curve that is mostly at the mean of zero with local excursions where non-zero outputs have been observed. Without being able to predict the true underlying objective function well, the BO algorithm has little chance of being able to find the global optimum of the problem unless the initial inputs start near it.

In practice, we often get only one shot at running the BO algorithm, because of a limited computing budget on the expensive simulator. So it is usually the case that we do not know if the BO algorithm was able to find the global solution. As we develop methodology for BO, we typically work with simpler examples that can be

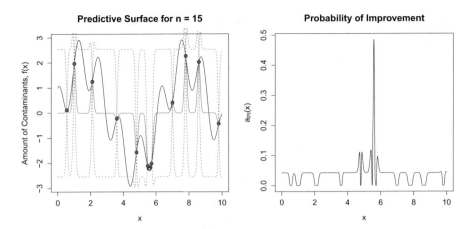

Fig. 3.6 The corresponding predictive surface (left) and PI surface (right) given 15 data points for the drilling example

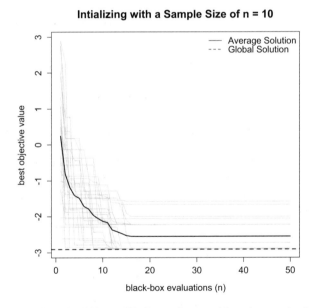

Fig. 3.7 The results of running 30 Monte Carlo experiments with random starting input sets drawn from a LHD of size $n = 10$

run quickly and repeatedly, and thus we can run experiments with multiple different initial starting inputs to see how well the BO algorithm performs, and how robust the solution to the problem is. A typical visual summary of multiple runs of the BO algorithm is to plot the best objective value found as a function of the number of input evaluations for each run. For example, Fig. 3.7 shows the trajectories of the best objective values found (grey lines) for 30 Monte Carlo experiments (i.e.,

Fig. 3.8 The results of
running 30 Monte Carlo
experiments with random
starting input sets drawn from
a LHD of size $n = 15$

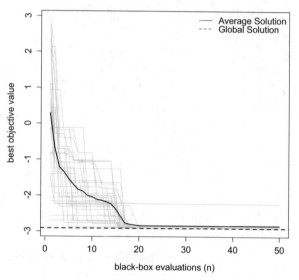

rerunning the BO algorithm with different initial input sets of size $n = 10$). The
average trajectory over the 30 Monte Carlo experiments is shown in black, and
we can see that on average the PI acquisition function does not find the global
solution to the problem. Some of the individual trajectories from the 30 Monte Carlo
experiments do find the global solution, however, these are instances where the
initial input sets contain a point close to the global solution. This figure demonstrates
the large amount of variability that can result from using the PI acquisition function,
even on a relatively simple one-dimensional problem.

In the case of the drilling problem, it turns out that choosing an initial set of inputs
from a LHD of size $n = 15$ leads to far better solutions to the problem (Fig. 3.8).
The reasoning for this is twofold. First, selecting a slightly larger initial LHS leads
to better space-filling which in turn gives a higher likelihood of starting with a point
near the global minimum. Second, the addition of slightly more points also leads
to better prediction of the true underlying contamination function by our Gaussian
process.

So why choose to use the PI acquisition function? Well it turns out it is not a good
choice of acquisition function when the objective function has lots of local minima,
however, when the objective function is relatively smooth, or with few local minima,
the PI acquisition can be very fast to locate the global solution to the optimization
problem. This good performance is typically tied to how well the Gaussian process
model does at actually predicting the objective function.

Lastly, recall that in Sect. 3.2 that an acquisition function can be thought of as the
maximization of a utility function. It turns out that some acquisition functions can be
interpreted in the framework of Bayesian decision theory (Berger 1985), where we
evaluate the expected utility associated with evaluating f at input x. From a decision

theoretic point-of-view, we should select the input that yields the highest expected utility. It turns out that the probability of improvement can also be motivated in this way. Consider the following utility function

$$u(x) = \begin{cases} 1 & \text{if } Y(x) \le f_{\min}^n \\ 0 & \text{otherwise} \end{cases}. \tag{3.5}$$

Now, under the given utility function $u(x)$, we can derive our acquisition function as a function of the expected utility gain for choosing input x, i.e.,

$$a_{\mathrm{PI}}(x) = \mathbb{E}\{u(x)|x, D_n\}$$

$$= 1 \times \int_{-\infty}^{f_{\min}^n} N(Y(x); \mu_n(x), \sigma_n(x))dY(x)$$

$$+ 0 \times \int_{f_{\min}^n}^{\infty} N(Y(x); \mu_n(x), \sigma_n(x))dY(x) \tag{3.6}$$

$$= \Phi\left(\frac{f_{\min}^n - \mu_n(x)}{\sigma_n(x)}\right)$$

where, again, $\mu_n(x)$ and $\sigma_n(x)$ are the mean and standard deviation of the predictive distribution of $Y(x)$, and $\Phi(\cdot)$ is the standard normal cdf. Thus, the input with the highest probability of improvement is the same as the input with the maximum expected utility.

3.3.2 Expected Improvement

Originally introduced in the computer modeling literature (Jones et al. 1998), the expected improvement (EI) acquisition function has become one of the most famous, and probably most used, acquisition functions in BO. Jones et al. (1998) defined the improvement statistic at a proposed input x to be $I(x) = \max_x\{0, f_{\min}^n - Y(x)\}$ where, after n runs of the computer model, $f_{\min}^n = \min\{f(x_1), \ldots, f(x_n)\}$ is the current minimum value observed. Since the proposed input x has not yet been observed, $Y(x)$ is unknown and can be thought of as a random variable where, conditional on the observed inputs $x_1, \ldots, x_n$, it can be modeled using a Gaussian process. Similarly, $I(x)$ will also be a random variable.

Intuitively, this improvement function makes a lot of sense. If the value of $Y(x)$ is greater than the current minimum value observed, $f_{\min}^n$, then we have not found a better solution and the improvement at that input should be zero. On the other hand, if $Y(x)$ is less than $f_{\min}^n$, then the improvement in our solution is equal to the amount that we have decreased the current observed minimum value. Put a different way, the improvement function, $I(x)$, tells us the utility of choosing input x as the next point to evaluate. Thus, we can think of $I(x)$ as a utility function, and from a Bayesian decision theoretic point-of-view we can maximize this utility function in order to select the input that gives us the maximum expected utility gain. In the notation of decision theory, let us rewrite $I(x)$ as the utility function $u(x)$ where we define $u(x)$ as follows

$$u(x) = \begin{cases} f_{\min}^n - Y(x) & \text{if } Y(x) \le f_{\min}^n \\ 0 & \text{otherwise} \end{cases}. \tag{3.7}$$

From the decision theoretic point-of-view, the EI acquisition function can be derived as the expectation of the utility function. Calculating the expectation is not hard, but it does require a bit of calculus. Let $z \sim N(0, 1)$ be a standard normal random variable, and define $v = (f_{\min}^n - \mu_n(x))/\sigma_n(x)$. Again, since we treat $Y(x)$ as coming from a Gaussian process then, conditional on a particular parameterization of the Gaussian process, the EI acquisition function is available in closed form as

$$\begin{aligned} a_{EI}(x) &= \mathbb{E}\{u(x)|x, D_n\} \\ &= \mathbb{E}\{(f_{\min}^n - Y(x))\mathbb{1}_{\{Y(x) < f_{\min}^n\}}(x)\} \\ &= \int_{-\infty}^{\infty} (f_{\min}^n - Y(x))\mathbb{1}_{\{Y(x) < f_{\min}^n\}}(x) \times N(Y(x); \mu_n(x), \sigma_n(x))dY(x) \\ &= \int_{-\infty}^{f_{\min}^n} (f_{\min}^n - Y(x)) \times N(Y(x); \mu_n(x), \sigma_n(x))dY(x) \\ &= \int_{-\infty}^{v} (f_{\min}^n - (\mu_n(x) + \sigma_n(x)z)) \times N(z; 0, 1)dz \\ &= \int_{-\infty}^{v} (f_{\min}^n - \mu_n(x) - \sigma_n(x)z) \times N(z; 0, 1)dz \\ &= (f_{\min}^n - \mu_n(x))\int_{-\infty}^{v} N(z; 0, 1)dz - \sigma_n(x)\int_{-\infty}^{v} z \times N(z; 0, 1)dz \\ &= (f_{\min}^n - \mu_n(x))\Phi\left(\frac{f_{\min}^n - \mu_n(x)}{\sigma_n(x)}\right) + \sigma_n(x)\phi\left(\frac{f_{\min}^n - \mu_n(x)}{\sigma_n(x)}\right) \end{aligned}$$
$$\tag{3.8}$$

where $\mathbb{1}_{\{\cdot\}}$ is the usual 0-1 indicator function, and $\mu_n(x)$ and $\sigma_n(x)$ are the mean and standard deviation of the predictive distribution of $Y(x)$, and $\Phi(\cdot)$ and $\phi(\cdot)$ are the standard normal cdf and pdf, respectively. Notice that $\Phi(\cdot)$ term in (3.8) is actually the PI acquisition function weighted by the amount that the predictive mean, $\mu_n(x)$, is below the current observed minimum value $f_{\min}^n$. When $\mu_n(x)$ is far below $f_{\min}^n$ the EI acquisition will tend to place high weight on exploitation (i.e., local search). The other term in the EI acquisition function, $\phi(\cdot)$, is weighted by the predictive variance and controls the level of global search of the algorithm. Inputs with high predictive variance will place higher value on searching globally. And so, the acquisition function in (3.8) naturally provides a combined measure of how promising a candidate point is, that trades off between local search ($\mu_n(x)$ below $f_{\min}$) and global search (large $\sigma_n(x)$).

Now recall again the improvement function $I(x) = \max_x\{0, f_{\min}^n - Y(x)\}$. Without the notion of Bayesian decision theory, the EI acquisition function arises somewhat organically from the notion that it just makes sense to choose new inputs to evaluate that have the highest expected improvement over other inputs. Statisticians tend to think about things in terms of averages, and so, an input that improves upon the current best solution is great, but an input that on average improves upon the current best solution is even better. Thus, it just feels natural to take the expected value of the improvement statistic, which again, yields

$$\mathbb{E}[I(x)] = (f_{\min}^n - \mu_n(x))\Phi\left(\frac{f_{\min}^n - \mu_n(x)}{\sigma_n(x)}\right) + \sigma_n(x)\phi\left(\frac{f_{\min}^n - \mu_n(x)}{\sigma_n(x)}\right).$$
(3.9)

Figure 3.9 provides a graphical display of the components of EI. The green Gaussian curve shows the predictive density at one possible point, x_1, and the blue Gaussian curve shows the predictive density at a second point, x_2. The height of the density curves at $f_{\min}^n$ is $\phi\left(\frac{f_{\min}^n - \mu_n(x_1)}{\sigma_n(x_1)}\right)$ and $\phi\left(\frac{f_{\min}^n - \mu_n(x_2)}{\sigma_n(x_2)}\right)$. The shaded regions

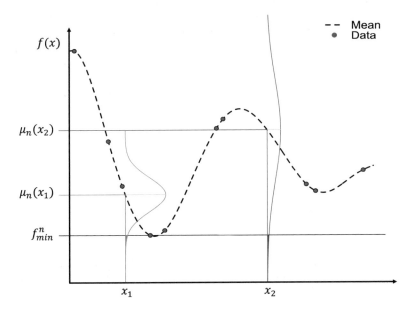

Fig. 3.9 An example of the components of Expected Improvement

represent $\Phi\left(\frac{f_{\min}^n - \mu_n(x_1)}{\sigma_n(x_1)}\right)$ and $\Phi\left(\frac{f_{\min}^n - \mu_n(x_2)}{\sigma_n(x_2)}\right)$, which is also the PI acquisition function. Point x_1 is favored by the first term, as it has a higher probability of positive improvement. Point x_2 will get more weight from the second term, as $\sigma_n(x_2) > \sigma_n(x_1)$.

Example To better understand what the EI acquisition function is doing in practice, let us consider the following one-dimensional example. Treating the objective function like a black-box function, we wish to minimize the following objective function

$$f(x) = -(1.4 - 3x)\sin(18x)$$

where the input $x \in [0, 1.2]$. The global solution to the problem is at $x = 0.96609$. Now imagine that you start with a LHS of ten inputs to initialize the BO algorithm, and that your Gaussian process surrogate model leads to the following predictive surface in Fig. 3.10. The choice of the next input to evaluate depends upon the EI acquisition function, and so we choose the input that maximizes the EI surface in Fig. 3.10. Clearly one side of the input space (roughly $0 \le x \le 0.15$) has a higher chance for expected improvement, however, there is a small blip of expected improvement around $x = 0.9$. The reasoning for the shape of the EI surface is simple. Given the locations of the evaluated data (Fig. 3.11), one data point falls extremely close to a very low local minimum, while two data points straddle the global minimum. The point near the local minimum is judged to be the best current solution and so points around it are given high expected improvement. On the other hand, the distance between the two points straddling the global minimum is just far enough to give the predictive surface high variability in that area. Remember that the EI acquisition function naturally balances the exploitation versus exploration trade-

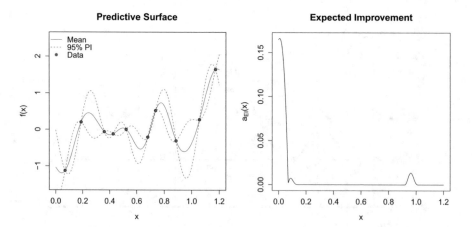

Fig. 3.10 The corresponding predictive surface (left) and EI surface (right) given 10 data points

True Objective Function

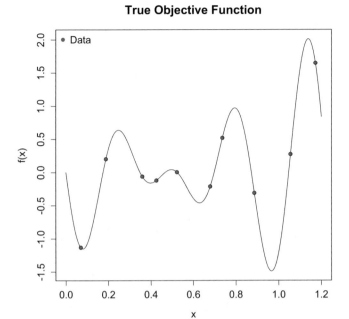

Fig. 3.11 The objective function $f(x) = -(1.4 - 3x)\sin(18x)$ with 10 evaluated inputs

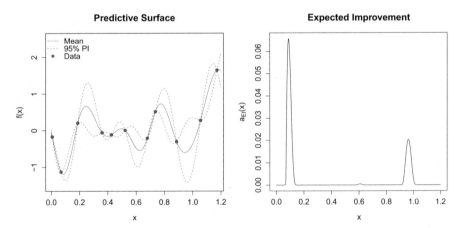

Fig. 3.12 The corresponding predictive surface (left) and EI surface (right) given 11 data points

off, and so areas with low predicted mean values or areas with high uncertainty are viable options for the BO algorithm to want to search.

Based on the EI surface, the best next input to evaluate is at the edge of the left side of the input space. Although this is the correct choice of input given the current shape of the EI surface, it turns out to be a poor choice, as we can see in Fig. 3.12,

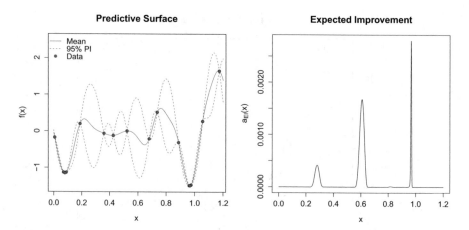

Fig. 3.13 The corresponding predictive surface (left) and EI surface (right) after evaluating 20 inputs

since the evaluated point is much higher than the current observed minimum value. We do not gain a better observed minimum value, but we do gain knowledge about the shape of the true underlying objective function. This is reflected in both the increase in accuracy of the predictive surface, as well as the updated EI surface. Now the large peak in the EI surface in Fig. 3.12 has decreased substantially, and the peak that once was once a small blip has grown. Again, the EI acquisition function balances exploring the input space versus exploiting it, and so areas of promise or high uncertainty are reflected in the EI surface.

If we were to continue to evaluate nine more inputs sequentially, then the BO algorithm would end up favoring jumping between the promising local minimum and the area of the global minimum. As we can see from Fig. 3.13, by the twentieth data point, the BO algorithm has effectively only evaluated points at the local and global minimum, but now that the area around the global minimum has been evaluated, the EI surface is drastically different than when we started. The EI surface now favors the right side of the input space, and smaller expected improvement blips in the EI surface appear on the left side of the input space where variability is still high. Continuing to run the BO algorithm on this problem, the search would continue in similar fashion where the EI acquisition function would guide the search to areas of high uncertainty eventually.

Example One-dimensional illustrations are useful for their ease of understanding, but let us now turn our attention to a slightly more complicated two-dimensional example. We wish to minimize the following objective function

$$f(x_1, x_2) = -(\cos((x_1 - 0.1)x_2))^2 - x_1 \sin(3x_1 + x_2),$$

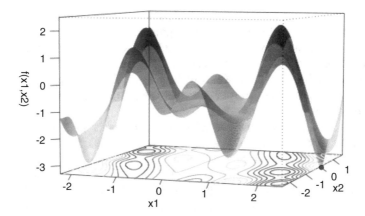

Fig. 3.14 The objective function $f(x_1, x_2) = -(\cos((x_1 - 0.1)x_2))^2 - x_1 \sin(3x_1 + x_2)$. The input space contains many local minima, but there is only one global minimum at the edge of the input space (blue point)

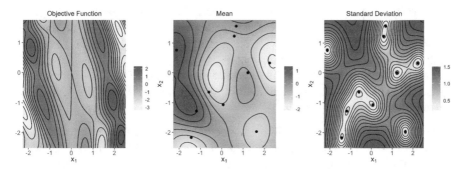

Fig. 3.15 A view of the objective function surface, and the predictive mean, $\mu_n(x)$, and standard deviation, $\sigma_n(x)$, surfaces after ten input evaluations. Here the inputs evaluated are shown as black points

where $-2.25 \leq x_1 \leq 2.5$ and $-2.25 \leq x_2 \leq 1.75$. The objective function is a smooth non-linear surface (Fig. 3.14) with a global minimum at the edge of the input space at $x_1 = 2.50$, $x_2 = 0.063$. Searching for the global solution is tough because there are several local minima to get trapped in, and moreover, many of the local minima have objective function values close the value of the global minimum.

Now, let us assume that we initialize the BO algorithm a LHS of size $n = 10$. Recall that both the PI and EI acquisition functions heavily depend upon both the predictive mean, $\mu_n(x)$, and variance, $\sigma_n^2(x)$, estimates from our Gaussian process surrogate model. Figure 3.15 gives us an idea of how well the Gaussian process model is doing at emulating the true objective function surface. With only ten points, the predictive mean surface captures some of the features of the objective function, and overall is not a terrible first approximation to the objective function. Remember that as the search for the global solution progresses, our knowledge of the objective function (i.e., evaluated inputs) will continuously be updated and in turn so will our estimate of the predictive mean surface. Likewise, as expected, our uncertainty around observed inputs should be low while areas of the input space less explored should still contain high amounts of uncertainty, which is exactly what we see in the plot of the standard deviation surface.

Both the PI and EI acquisition function use the exact same predictive mean, $\mu_n(x)$, and variance, $\sigma_n^2(x)$, estimates, however, clearly the two acquisition functions do not need to be in agreement with the choice of best next input to evaluate. In fact, the two acquisition functions may yield surfaces that are entirely dissimilar, or surfaces that look similar but that still place higher importance in different regions of the input space. For example, in Fig. 3.16 the surfaces under the PI and EI acquisition functions look resoundingly similar, however, the two acquisition functions clearly choose two very different best next inputs to evaluate (green and light blue point, respectively). As mentioned before, the PI acquisition function tends to lead to a greedy BO algorithm and so in this example the PI acquisition functions picks the green point very close to the global solution simply because there is an observed input close to there. On the other hand, the EI acquisition function favors searching in an area that has both a promising predictive mean value (i.e., potentially lower than the best current observed value $f_{\min}^n$) and a substantial amount of uncertainty around it. Again this is a key feature of the EI acquisition function; balancing exploitation and exploration by appropriately weighting the decision to search either locally or globally.

Example Let us revisit the drilling for contaminants problem of Sect. 3.1. In Sect. 3.3.1, we observed less than desirable behavior of the PI acquisition function. The PI acquisition function was quick to get stuck in areas of local minima if there was no data point reasonably close to the global minimum. Given the ability of the EI acquisition function to trade off between local and global search, it seems reasonable that a BO algorithm using this acquisition function should be able to

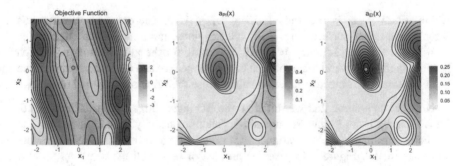

Fig. 3.16 The best next inputs (green and light blue points) to be evaluated under the PI and EI acquisition functions

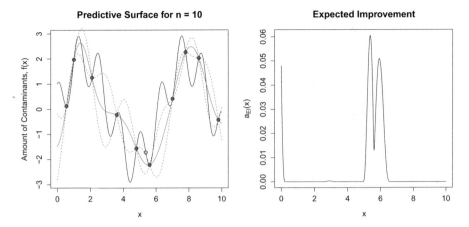

Fig. 3.17 The corresponding predictive surface (left) and EI surface (right) given 10 data points for the drilling example. The blue point corresponds to the next input to evaluate based on the EI surface

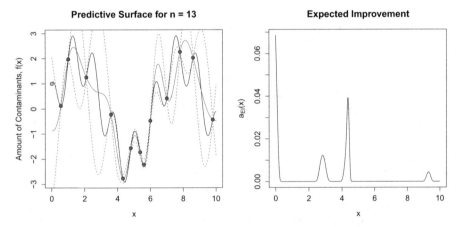

Fig. 3.18 The corresponding predictive surface (left) and EI surface (right) given 13 data points for the drilling example. The blue point corresponds to the next input to evaluate based on the EI surface

avoid this type of greedy behavior. To test this hypothesis, we start with the same 10 inputs from the contaminants example in Sect. 3.3.1, and similarly, sequentially choose five more inputs based on the EI acquisition function. The EI surface (Fig. 3.17) looks somewhat similar to the PI surface (Fig. 3.4), and so the next input selected is around the same location in the local mode. However, where the EI and PI acquisition functions diverge is in their selection of the next inputs.

After an additional three inputs have been sequentially chosen and evaluated, we see that the EI acquisition function (Fig. 3.18) is able to avoid getting stuck

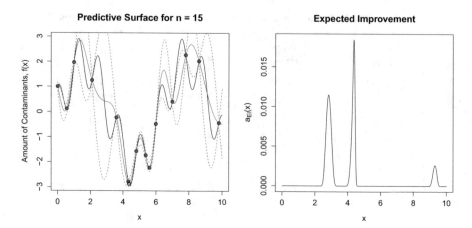

Fig. 3.19 The corresponding predictive surface (left) and EI surface (right) given 15 data points for the drilling example. The blue point corresponds to the next input to evaluate based on the EI surface

in the local minimum of the function that it initially began to explore, which was not the case for the PI acquisition function (Fig. 3.5). In the case of the PI acquisition function, the PI surface remained extremely peaked in the area of the local minimum. In contrast, the EI surface is multi-modal, with all of the peaks of the EI surface centered over areas where the predictive mean is either lower than f_{min} or where the predictive variance is very high. The contribution of the predictive variance to the EI acquisition function is what allows it to be able to escape local minima.

At the fifteenth iteration of the algorithm, the EI surface becomes increasingly peaked in the location of the global solution (Fig. 3.19). However, the predictive uncertainty in areas less explored will always creep back up in the EI surface. Areas of the input space with small expected improvement will naturally increase in relative EI as the predictive variance in those regions remains large as compared to areas that have been frequently more explored.

To finish the comparison, we run 30 Monto Carlo experiments where we initialize the BO algorithm with a LHS of size $n = 10$, and another 30 Monto Carlo experiments where initialize with a LHS of size $n = 15$. In both cases, we record the trajectories of the best objective values found (grey lines in Fig. 3.20) from each of the Monte Carlo experiments. As expected, the EI acquisition function does a better job as compared to the PI acquisition function, although even the EI acquisition function is not perfect at finding the global solution of such a wiggly function. Again, we see the behavior that an increase in initial sample size (from 10 to 15) can benefit the acquisition function.

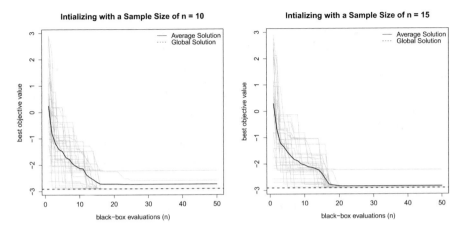

Fig. 3.20 The results of running 30 Monte Carlo experiments with random starting input sets drawn from a LHD of size $n = 10$ or $n = 15$

3.3.3 Lower Confidence Bound

Not all acquisition functions can be interpreted as the derivation of an expected utility function under the Bayesian decision theoretic framework. Instead, sometimes acquisition functions are simply functions that that have been cleverly constructed to balance the exploration-exploitation trade-off while searching for the global solution to the optimization problem. One such way is the confidence bound criteria introduced in Cox and John (1997), and later adapted by Srinivas et al. (2010), known as the lower confidence bound (LCB) acquisition function. The LCB acquisition function is a simple linear combination of the predictive mean, $\mu_n(x)$ and variance, $\sigma_n(x)$, of the surrogate model, and can be expressed as follows:

$$a_{\text{LCB}}(x; \beta) = -\mu_n(x) + \beta\sigma_n(x) \tag{3.10}$$

where $\beta \geq 0$ is a tuning parameter. The exploration-exploitation trade-off is naturally balanced by the acquisition function in (3.10) since it favors both inputs, x, where the objective function is uncertain (large $\sigma_n(x)$) and inputs where the minimum is promising (optimal $\mu_n(x)$).

Recall that we think of the inner optimization problem within the BO algorithm as a maximization step since we view the acquisition function as a utility function. If we forgo this notion for a moment, and instead think of the acquisition function as a minimization problem, then it makes sense to write the LCB acquisition function as

$$\text{LCB}(x; \beta) = \mu_n(x) - \beta\sigma_n(x). \tag{3.11}$$

Here we drop the $a(\cdot)$ notation to stress the fact we are not maximizing an acquisition function. The appeal of viewing the acquisition function in this way is that it is clear to see what the underlying form of the function is. What the acquisition function suggests is that we should choose points that minimize the lower quantiles of the predictive distribution of $Y(x)$. In fact, for appropriate choices of β, this can be viewed as choosing the lowest lower bound on the $(1 - \alpha)\%$ probability (or credible) interval for $\mu_n(x)$. And so, use of the LCB acquisition function can proceed by either maximizing (3.10) or minimizing (3.11). Although the focus of this book is on minimization of the objective function, it is worth mentioning that there exists the analogous upper confidence bound acquisition function for maximizing the objective function, i.e.,

$$a_{\mathrm{UCB}}(x; \beta) = \mu_n(x) + \beta\sigma_n(x). \tag{3.12}$$

Returning to the LCB aquisition function in (3.10), the tuning parameter β is left as a free parameter set at the user's discretion. However, a typical thing to do is to allow β to change over time such that, at every iteration of the of the BO algorithm, we choose the input that maximizes the LCB acquisition function

$$a_{\mathrm{LCB}}(x; \beta_n) = -\mu_n(x) + \beta_n\sigma_n(x). \tag{3.13}$$

As the optimization progresses, this sequence of tuning parameters, β_n, also helps to further the balance between local and global search. Larger β_n lead to more global searches, while small β_n favor more local searches. In fact, setting $\beta_n = 0$ reduces to simply minimizing the predictive mean surface as if it were known exactly without uncertainty, i.e., greedily selecting x_n such that

$$x_n = \operatorname*{argmin}_{x \in \mathcal{X}} \mu_{n-1}(x). \tag{3.14}$$

Strong theoretical results exists for the LCB acquisition function under different parameterizations of β_n. For example, under certain conditions (Srinivas et al. 2010), a lower-bound on the convergence rate for the optimization can be derived. However, as pointed out in Gramacy (2020), in practice such theoretically automatic selections for β_n remain difficult and unwieldy for practitioners.

Example Let us revisit the example from the previous section where we wish to minimize the following objective function

$$f(x) = -(1.4 - 3x)\sin(18x)$$

over the input space $x \in [0, 1.2]$. Here the focus of this example is to try to understand the behavior of the LCB acquisition function. As mentioned previously, we find that working with the form of the function in (3.11) is more intuitive visually, and so here we focus on minimizing the function $\mathrm{LCB}(x; \beta)$ rather than maximizing the acquisition function $a_{\mathrm{LCB}}(x; \beta)$. Again, assume we start with a

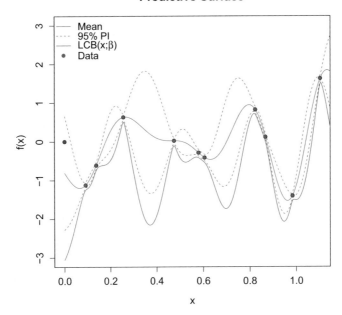

Fig. 3.21 The predictive surface and LCB function given 10 data points. The blue point is the best next input that would be evaluated by the BO algorithm

LHS of size $n = 10$ to initialize the BO algorithm. Now unlike the PI and EI acquisition functions, proceeding to use the LCB function requires some thought about specifying the tuning parameter β. Small values of β will promote local search, while large values of β will make the BO algorithm favor exploring globally in areas with high uncertainty, however, note that small and large are terms we use loosely as what is considered small and large in one problem may not be the same in another. In practice, a schedule of values for β as the BO algorithm progresses is generally preferred, however, here for the sake of example we simply set $\beta = 3$. Figure 3.21 highlights the relationship between components $\mu_n(x)$ and $\sigma_n(x)$, and how they affect the LCB function. The LCB function mimics the shape of the lower limit of the 95% probability interval for the mean prediction and is a scaled version of it. Here, β plays the role of the scaling factor where large β increases the distance between the lower limit of the 95% probability interval (red dashed line) and the LCB function (blue solid line). Conversely, small values of β shrink the LCB function surface towards the predictive mean surface. Similar to the results of the EI acquisition function in the previous section, the best next input (blue point) to be selected by the BO algorithm is at the far left side of the input space where the predictive mean is low and the uncertainty around the mean is also high.

After we have evaluated the next input we see that the uncertainty at the edge of the input space has reduced greatly (Fig. 3.22). At this point, it should be more

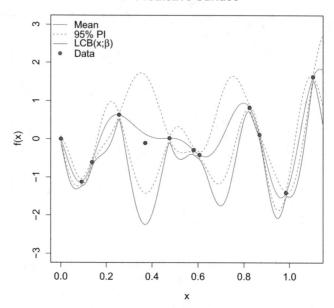

Fig. 3.22 The predictive surface and LCB function given 11 data points. The blue point is the best next input that would be evaluated by the BO algorithm

apparent that the LCB function naturally balances exploration versus exploitation as well. The best next input to be chosen is selected in a region of the input space where the predictive mean is nowhere near as low or promising as other regions, however, the the large amount of uncertainty in the region leads the LCB function to believe that it is a promising area to explore in. It is worth noting that areas of either high uncertainty or low predicted mean values are both given some importance to search at by the LCB function, which again helps to balance local versus global search.

As the BO algorithm search progresses, the LCB function surface becomes highly multimodal (Fig. 3.23). This multimodal feature is a consequence of using the Gaussian process model as our surrogate model. At observed inputs, the predictive uncertainty shrinks towards zero (or is exactly zero in the case where a nugget is not used; see Sect. 2.1) and so the corresponding probability interval reflects this fact in its ballooning shape. Since the LCB function is a weighted version of the lower limit of the probability interval, this feature will also be reflected in the LCB function surface.

Lastly, the choice of the value of $\beta = 3$ was completely arbitrary in this example, however, in practice careful consideration should be given to selecting the appropriate value, or values, of β in the acquisition function. Different values of β can greatly influence the performance of the BO algorithm. For example, let us consider re-running the BO algorithm on this objective function with varying values of β. Here we can let β equal either 0, 0.5, 1, 2, or 4, and repeat solving

Predictive Surface

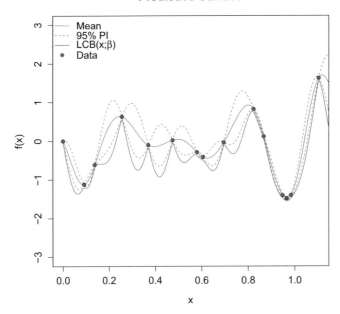

Fig. 3.23 The predictive surface and LCB function given 20 data points. The blue point is the best next input that would be evaluated by the BO algorithm

this minimization problem using 30 different Monte Carlo experiments. Here we will initialize the BO algorithm with a LHS of $n = 10$, and sequentially select 20 more inputs for a total budget of 30 input evaluations. The trajectories of the average best solution over the 30 Monte Carlo experiments for each β are plotted in Fig. 3.24. It is clear that the average performance of the BO algorithm is directly tied to the choice of β. Set β too small, and the BO algorithm will get stuck in local modes, while setting β too large can result in over exploration of the input space, and consequently a BO algorithm that is slower to converge to the global solution.

Example We revisit one last time the drilling for contaminants problem of Sect. 3.1. Having previously explored the BO algorithm for this problem under the PI and EI acquisition functions, we shall solve it once again but this time using the LCB function in (3.11). We start again with the same 10 inputs from the contaminants example in Sect. 3.2, and similarly, sequentially choose five more inputs based on the LCB function with $\beta = 3$. Unlike the PI and EI acquisition functions, the LCB function predicts that the next best input is at the far edge of the input space (Fig. 3.25). This choice makes a lot of sense given the nature of the probability intervals associated with Gaussian process prediction. At the edges of the input space, the uncertainty associated with our Gaussian process prediction is high until a point has been evaluated along the edge, and the LCB function favors picking points where the lower limit of the quantiles of the predictive distribution of

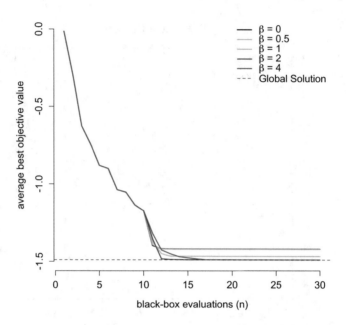

Fig. 3.24 The results of running 30 Monte Carlo experiments using the LCB function for various values of β

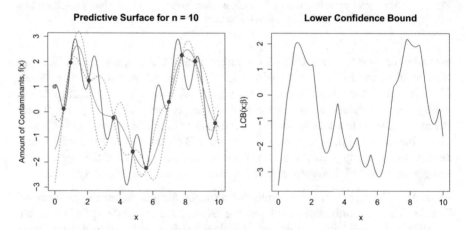

Fig. 3.25 The corresponding predictive surface (left) and LCB surface (right) given 10 data points for the drilling example

$Y(x)$ are at their lowest. The LCB function surface (Fig. 3.25) also reflects this fact given that the LCB function places high importance on both ends of the input space (i.e., low LCB$(x; \beta)$ values), as well as high importance searching around where the lowest observed input is.

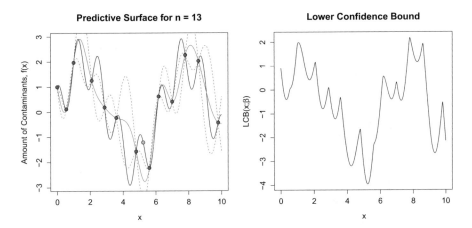

Fig. 3.26 The corresponding predictive surface (left) and LCB surface (right) given 13 data points for the drilling example

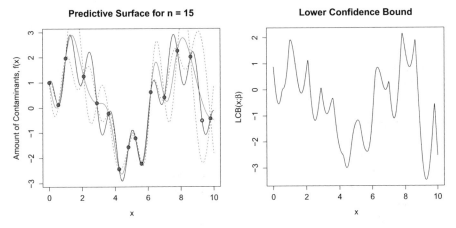

Fig. 3.27 The corresponding predictive surface (left) and LCB surface (right) given 15 data points for the drilling example

After an additional three inputs have been selected sequentially, we see that the uncertainty in our predictive distribution starts to diminish (Fig. 3.26) and so the lowest quantile values of the predictive distribution appear around where the global solution is. However, as no further input evaluations have occurred at the right side of the input space, the LCB function still views it as a somewhat promising location to search in.

By the fifteenth input evaluation, the uncertainty at the right side of the input space dwarfs all other uncertainty (Fig. 3.27) and so the LCB function selects the best next input to evaluate over there since its corresponding quantile of the

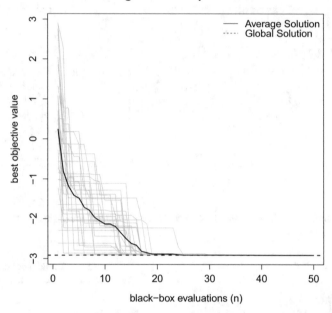

Fig. 3.28 The results of running 30 Monte Carlo experiments with random starting input sets drawn from a LHD of size $n = 10$

predictive distribution is also lowest. Note that at this point, the BO algorithm has not got stuck searching in a local mode as was the case when using the PI acquisition function. Like the EI acquisition function, the LCB function is capable of balancing exploration versus exploitation. At this point, we can see that the BO algorithm has not yet honed in on the global solution, but it is clear to see that it is well on its way to finding it.

To finish the comparison, we run 30 Monto Carlo experiments where we initialize the BO algorithm with a LHS of size $n = 10$. Here, we record the trajectories of the best objective values found (grey lines in Fig. 3.28) from each of the Monte Carlo experiments. Again, given no hypothetical access to any subject matter expertise about drilling for contaminants, we choose to fix $\beta = 3$. Somewhat surprisingly, the average solution to the problem, under the LCB function, is much faster to converge to the global solution of the problem as compared to the PI (Fig. 3.7) and EI (Fig. 3.20) acquisition functions at an initial input sample size of 10.

3.3.4 Other Acquisition Functions

At this point, we have explored three different acquisition functions (PI, EI, and LCB) for guiding the BO algorithm, however, they are by no means the only acquisition functions that exist. Each of these three acquisition functions come with their own lists of pros and cons, yet quite frankly, no one of these three acquisition functions uniformly dominates the other ones. The success (and failure) of a given acquisition function is highly driven by the attributes of the objective function at hand. Not surprisingly, more and more acquisition functions are continuously being created given the increasing complexity and nature of optimizing black-box functions. In what follows, we briefly mention a few more well known acquisition functions for unconstrained BO.

Random Search

The development of novel acquisition functions is an increasing area of interest within the BO community. The goal of developing novel acquisition functions is to generate better functions for guiding the BO algorithm in search of the optimum. Here, better is defined by characteristics and metrics that improve the performance of the BO algorithm, such as speed (i.e., reducing computational time) of the BO algorithm or reducing the number of black-box evaluations to reach the global solution, to name a few. However, one of the most important, and often neglected, performance metrics that should be checked during acquisition function development is whether or not the proposed acquisition function provides a BO algorithm that returns a solution that is better than a random search. Here we define random search by the following acquisition function

$$a_{RS}(x) = c, \tag{3.15}$$

where $c \in \mathbb{R}$ is a constant. Any input $x \in X$ maximizes this acquisition function, and so the BO algorithm can proceed to sequentially search for the global solution to the minimization problem by simply picking a sequence of random inputs x to evaluate. In other words, if your total computational budget is N input evaluations, then take a random sample of size N from the input space to evaluate the objective function at, and then return the minimum objective function value as the best solution found. Remarkably, this random search concept can work quite well in low dimensional problems with even modest computational budgets. Now, this simple to implement and fast random search acquisition function may seem crude in nature, however it is a very useful acquisition function in that it serves as a sort of minimum benchmark that must be beat by other acquisition functions. Otherwise, why use them?

Thompson Sampling

Originally developed in Thompson (1933), Thompson sampling is a stochastic search algorithm that selects inputs to evaluate by maximizing the expected reward (or utility) at input x with respect to a randomly drawn sample from a posterior distribution of interest.

In our context of Bayesian optimization, Thompson sampling can be formulated as maximizing the following acquisition function:

$$a_{TS}(x) = -y_f(x) \tag{3.16}$$

where $y_f(x)$ is a draw from the posterior predictive distribution of the Gaussian process for the objective function f. As pointed out in Gramacy (2020), the TS acquisition function feels a lot like choosing the input that maximizes the negative predictive mean, $-\mu_n(x)$, of the Gaussian process, but with the added twist of a sense of uncertainty incorporation in the maximization step since we are sequentially maximizing an independent random draw of the predictive distribution at each iteration of the BO algorithm. Readers interested in learning more about Thompson sampling in the context of Bayesian optimization are advised to see Shahriari et al. (2016) and the sources within it.

Entropy Search

Introduced in Henning and Schuler (2012), the entropy search acquisition function guides the BO algorithm by choosing inputs to evaluate that would cause the largest average decrease in differential entropy. Here, for a random variable Z with probability density function g, whose support is the set $\mathcal{Z}$, the differential entropy $H(Z)$ is defined as

$$H(Z) = -\int_{\mathcal{Z}} g(z) \log g(z) dz. \tag{3.17}$$

Here, smaller differential entropy indicates less uncertainty in our random variable. Casting this problem in the notation of utility functions, we can write the decrease in differential entropy for having evaluated input x as

$$u(x) = H(x^*|D_n) - H(x^*|D_n, x, Y(x)) \tag{3.18}$$

where $x^* = \operatorname{argmin}_{x \in \mathcal{X}} f(x)$. With this understanding, $H(x^*|D_n)$ represents the entropy associated with the posterior predictive distribution of the Gaussian process at iteration n, while $H(x^*|D_n, x, Y(x))$ represents the entropy associated with the posterior predictive distribution of the Gaussian process at iteration $n+1$ if we were to evaluate input x. As we saw with previous acquisition functions in Sects. 3.3.1–3.3.2, we can use Bayesian decision theory to derive the acquisition function under the given utility function by evaluating the expected utility, i.e.,

$$a_{\mathrm{ES}}(x) = \mathbb{E}\{u(x)|x, D_n\}$$
$$= H(x^*|D_n) - \mathbb{E}\{H[x^*|D_n, x, Y(x))\}. \tag{3.19}$$

Here, the expectation is taken with respect to $Y(x)$. Unfortunately, evaluating this acquisition function is non-trivial as no closed form solution exists. Rather, a series of challenging approximation steps is required to maximize the entropy search acquisition function. Readers interested in a slightly less challenging to evaluate entropy based acquisition function are encouraged to see the work on the predictive entropy search acquisition function in Hernández-Lobato et al. (2014).

3.4 Sprinkler Computer Model

Up until now, all of the examples in this chapter have relied upon optimizing a known objective function that we artificially treated as a black-box function. To give an illustration of the performance of the acquisition functions on a real-world black-box computer experiment, we turn our attention to the sprinkler computer model introduced in Sect. 1.2.4. One of the objectives of the sprinkler computer model is to maximize the range at which the garden sprinkler can spray water given the eight different physical parameters $(x_1, \ldots, x_8)$ of the garden sprinkler (Fig. 1.6). Here, $x = (x_1, \ldots, x_8)^T \in \mathcal{X}$ constitutes the eight dimensional input space (Table 1.1) over which we wish to maximize the black-box objective function. In this chapter, we focus only on maximizing the range of the garden sprinkler irrespective of whatever the implied values of the water consumption and speed of the sprinkler rotation may be. In the next chapter, we consider the problem of maximizing the range of the garden sprinkler subject to the constraint that the water consumption does not exceed a specified value. Yet, another possibility would be to treat this problem as a multi-objective optimization problem, where one attempts to jointly optimize all three outputs; however, multi-objective problems are beyond the scope of this book. Like with the contaminant drilling problem in Sect. 3.1, we want to find a global maximum, so in the formulation of this book, we instead minimize the negative objective function in order to find the input that maximizes it.

Now, we shall solve for the minimum negative value of the range of the garden sprinkler using the probability of improvement (PI), expected improvement (EI), and lower confidence bound (LCB) acquisition functions, and shall compare and contrast their performances. Likewise, as a benchmark, we will solve the optimization problem using the random search acquisition function (Sect. 3.3) as well. For each acquisition function, we initialize the BO algorithm with the same starting LHS of size $n = 10$ inputs. We plan for a total computational budget of 100 input evaluations and so we will sequentially select an additional 90 inputs, based on the respective acquisition functions, to evaluate. In order to assess the robustness of the solutions of the BO algorithm, under the different acquisition functions, we shall repeat solving this optimization problem several times under

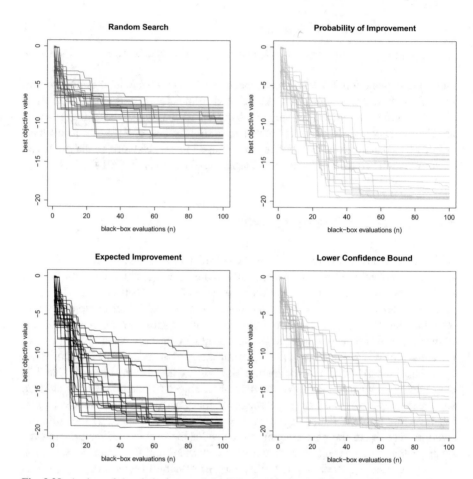

Fig. 3.29 A view of the performance of the BO algorithm, using the four different acquisition functions, for the 30 the Monte Carlo experiments. Here, each line represents the best value found over the search by the BO algorithm during a single run of the Monte Carlo experiment

different starting LHS. Here we shall set the number of Monte Carlo experiments to be 30. Figure 3.29 shows the results of the 30 Monte Carlo experiments for a given acquisition function.

Visually, it looks like the PI, EI, and LCB acquisition functions all have similar performance, and all three acquisition functions clearly outperform random search in general. For each acquisition function, taking the average of the solutions over the 30 Monte Carlo experiments reveals that the LCB acquisition function performed marginally better than the other acquisition functions (Fig. 3.30). Not surprisingly, the PI acquisition performed the worst at the beginning when compared to EI and LCB, which makes sense given the PI acquisition function's penchant for greedy search. Table 3.1 gives a numerical summary of the performance of the four acquisition functions. With the exception of random search, each acquisition function

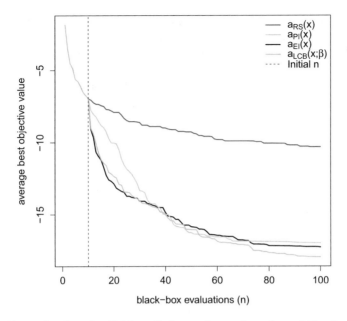

Fig. 3.30 The results of running 30 Monte Carlo experiments for each acquisition function. The plot shows the average best objective function values found over 100 black-box iterations

Table 3.1 The average, best, and worst solution found at the end of the 30 Monte Carlo experiments by each acquisition function

Acquisition function	Average final solution	Best final solution	Worst final solution
Random search	−10.32	−13.89	−7.54
Probability of improvement	−16.98	−19.66	−11.09
Expected improvement	−17.26	−19.68	−9.32
Lower confidence bound	−17.94	−19.68	−10.80

performed slightly better for a different metric of performance. For example, the LCB acquisition function has the lowest average final solution, however, PI had the least worst final solution amongst all of the acquisition functions.

Although not usually feasible in practice, this simulator is fast enough that we are able to evaluate a very large random sample of inputs. Evaluating a LHS of size $n = 1,000,000$ and $n = 10,000,000$ inputs results in global solutions of −16.86 and −17.68, respectively. It is striking that each of PI, EI, and LCB did better on average with 100 function evaluations than a LHS with 1,000,000 function evaluations. Increasing the size of the already rather large LHS by a factor of 10 only reduces the solution of -16.86 to -17.68. Quite amazingly, even with this extremely large (and impractical) LHS, random search did not find as low of a solution as any of the PI, EI, and LCB acquisition functions did using far fewer function evaluations. This result underscores the importance and usefulness of BO for efficiently searching the input space.

Chapter 4
Constrained Optimization

4.1 Constrained Bayesian Optimization

Constrained optimization is just the extension of unconstrained optimization to the case of having constraints. A constraint is a condition, equality, or inequality that must be satisfied in order for the solution to be valid. We can write the constrained optimization problem as

$$x^* = \operatorname*{argmin}_{x \in \mathcal{X}} f(x) \ \text{ subject to } \ c(x) \leq 0, \tag{4.1}$$

where $f : \mathcal{X} \to \mathbb{R}$ denotes a scalar-valued objective function for $\mathcal{X} \subset \mathbb{R}^d$ and $c : \mathcal{X} \to \mathbb{R}^m$ denotes a vector of m constraint functions. These constraints could be in the form of inequalities, $c_j(x) \leq 0$, equalities, $c_j(x) = 0$, or binary constraints represented with an indicator function, $c_j(x) = \mathbb{1}_{\{x \in \mathcal{X}_{c_j}\}}(x) = 0$.

Constrained optimization problems are typically more difficult than unconstrained problems because the constraints often operate at odds with the objective function, so trying to meet the constraint will drive the solution away from the global optimum. In simple cases, the optimum is not along a constraint boundary, in which case the problem is essentially an unconstrained problem. But in most cases, the optimum is on a boundary, with the unconstrained solution lying in a region that does not meet all the constraints. We refer to the subset of $\mathcal{X}$ that satisfies the constraints as the *valid* region. So one needs to be able to search for the best solution in the valid region, with the constraints taking first priority, and the objective function being second priority. More complex problems will have non-convex constraint boundaries, or even disconnected valid regions, both of which make searching quite difficult.

© The Author(s), under exclusive license to Springer Nature Switzerland AG 2021
T. Pourmohamad, H. K. H. Lee, *Bayesian Optimization with Application
to Computer Experiments*, SpringerBriefs in Statistics,
https://doi.org/10.1007/978-3-030-82458-7_4

Thus Bayesian optimization in the constrained case needs to learn both the objective function and the valid region. For both the objective and the constraints, exploitation needs to be balanced with exploration, although the interplay between the objective and the constraints needs to be taken into account. Exploration of the objective only needs to occur in the valid region, and exploration of the valid region only needs to occur for more optimal values of the objective. An efficient algorithm will be able to quickly hone in on the most promising parts of the space.

Again, constrained optimization is usually difficult because at least one of the constraints operates in opposition to the objective function, that is, they are negatively correlated. When the outputs are known or suspected to be correlated, it is common practice to use latent processes to induce a correlation structure between them (Sammel et al. 1997; Moustaki and Knott 2000, for example), however, this modeling choice comes at the cost of increased model complexity. Likewise, there is an increase in the time and computational burden of joint modeling as compared to independent modeling of the objective and constraint functions, however, as seen in Pourmohamad and Lee (2016), there can be significant gains in predictive accuracy and statistical coverage by the use of joint modeling when the objective and constraint functions are indeed correlated. Moreover, when the objective and constraint function are correlated, the shared information in modeling the functions jointly can lead to better model fits and prediction, which should lead to far fewer function evaluations of the expensive computer model in converging to the global solution to the optimization problem.

So why not simply use a joint model, such as a joint Gaussian process model (Wackernagel 2003), to model the objective and constraint functions jointly? After all, a joint Gaussian process model for the objective and constraint function will theoretically perform at least as well as using independent Gaussian processes models for each. The answer is two-fold. First off, if the objective and constraint function are not strongly correlated, then the added overhead and complexity in fitting the joint Gaussian process does not justify its use when, given the expensive nature of the computer model, computational speed of the BO algorithm is of utmost importance. Little is to be gained when there is not strong information to share across the models. Secondly, and quite frankly, independent Gaussian process surrogate models typically just work. Even without exploiting the joint information that exists for correlated outputs, independent Gaussian process surrogate models often do a fantastic job of predicting the objective and constraint functions well, and in a fraction of the time as compared to joint models. In the context of BO, it usually makes practical sense to use independent surrogate models for the objective and constraint functions.

With all of this in mind, generalizing the unconstrained BO algorithm (Algorithm 1 of Sect. 3.1) to the constrained case requires only a few additional pieces.

Here we describe the general formulation for constrained BO via the following algorithm:

Algorithm 2: The general constrained BO algorithm

Initialization:
    Start with an initial data set D_0
for $n = 1, \ldots, N$ **do**
    | Fit surrogate models for the objective and constraint functions;
    | Select $x_n = \text{argmax}_{x \in \mathcal{X}} a_{n-1}(x)$;
    | Evaluate f and c at x_n to obtain y_n;
    | Augment data $D_n = D_{n-1} \cup \{(x_n, y_n)\}$;
end
Return:
    $x^* = \text{argmin}_{x \in \mathcal{X}} f(x)$

Note that, in general, the only difference between the BO algorithm in the constrained case versus the unconstrained case is that a (either joint or independent) model must be specified for the constraint function, $c(x)$, as well as the objective function, $f(x)$. Otherwise, all other steps proceed similarly to the unconstrained case, i.e., start with an initial sample chosen from a space-filling design, and fit appropriate surrogate models that can be used to maximize an acquisition function for choosing the best next input at which to evaluate the computer model. This iterative algorithm repeats, updating the observed data after every iteration, until all computational budget has been exhausted. Different acquisition functions are needed in the presence of constraints, and that is the topic of the next section.

4.2   Choice of Acquisition Function

Just like the case of unconstrained Bayesian optimization, the acquisition function is the key to constrained Bayesian optimization. We introduce three different ways to think about the choice of an acquisition function here, although there are additional possibilities. These three are: (1) joint exploration of the objective function and the constraint functions, (2) focusing on staying inside the valid region, and (3) combining statistical modeling with numerical methods for effective exploration and exploitation. These approaches can overlap, and some examples of methods we present here are examples of more than one of these approaches.

One approach to constrained optimization is to choose an acquisition function that drives simultaneous learning of both the objective function and the constraint functions. By creating a good surrogate model for both f and c, one can find the global minimum using the surrogate model. In practice, the surrogate model for the objective function only needs to be accurate in the valid region, and the surrogate model for the constraints only needs to be accurate in regions of relatively lower objective function values. Two examples discussed in more detail in this chapter are Constrained Expected Improvement (Schonlau et al. 1998; Gardner et al. 2014) (Sect. 4.2.1) and Augmented Lagrangian methods (Gramacy et al. 2016) (Sect. 4.2.3). Additional examples include integrated expected conditional improvement (Gramacy and Lee 2011), expected volume minimization (Picheny 2014), and constrained BO for noisy experiments (Letham et al. 2019). As these approaches are learning both f and c, they tend to sample a relatively larger number of points near the constraint boundary but outside of the valid region, where f may be more desirable and c is close to being satisfied. These observations can help the search, although they will not ultimately be the optimum point because the constraints are not all satisfied.

A second approach is to focus on staying inside the valid region. This approach is motivated by interior point methods from numerical optimization, where the acquisition function is chosen so that if you have a starting point inside the valid region, the search attempts to remain inside the valid region, driving toward the boundary without crossing it. A prime example would be Barrier Methods (Pourmohamad and Lee 2021) (Sect. 4.2.4). Another approach to focusing on the valid region is Asymmetric Entropy (Lindberg and Lee 2015) (Sect. 4.2.2). The formulation of the constrained optimization problem in (4.1) basically assumes that the objective function f and the constraint function c can be evaluated for all $x \in \mathcal{X}$. In some cases, the computer model might not return any value when x is not in the valid region. For example, some inputs x might lead to trying to take the logarithm of a negative number somewhere in the code, or a matrix may become numerically singular in double precision and thus become not invertible. The problems may occur deep in the calculations, and a significant amount of computing may be necessary before discovering the issue. This situation is a case where the optimum must be found within the domain for which the computer model is able to return values. Outside of this valid region, the code fails to run and does not return a value. Thus c is observable as a binary variable, and f is only observable when $c = 0$. In the standard case of (4.1), every new observation provides some information toward optimization. However, in the case of code that does not run outside the valid region, information is only gained for runs inside the valid region. When the computer model fails to complete a run, the time spent on that attempted run is wasted. Thus it is much more important in this case to keep as many runs as possible inside the valid region. This class of methods aims to find the constrained optimum while limiting the number of observations with $c > 0$.

A third approach is based on hybrid optimization algorithms. These combine direct numerical optimization methods with statistical surrogate modeling. The idea is that the surrogate model contributes good exploration, and the numerical method contributes good exploitation. The two methods are combined in a way that balances exploration and exploitation, in order to efficiently find the global constrained minimum. Augmented Lagrangian (Sect. 4.2.3) and Barrier Methods (Sect. 4.2.4) are discussed in this chapter. Additional examples include the slack-variable augmented Lagrangian (Picheny et al. 2016), the ADMM algorithm for solving an augmented Lagrangian relaxation (Ariafar et al. 2019), Filter Methods

(Pourmohamad and Lee 2020), and scalable constrained BO based on trust regions (Eriksson and Poloczek 2021). Direct numerical optimization algorithms can be very efficient at finding a local optimum, and thus work quite well when run with a starting point in the domain of attraction of the global optimum, but can often become stuck in a local mode when started elsewhere. Statistical surrogate models can efficiently approximate the full surface. By using the surrogate model to guide the numerical algorithm, the hybrid approach can efficiently identify promising regions and find the local optimum of each, thus finding the global optimum as the best of the local optima. Some hybrid approaches work by iterating between the surrogate model and the numerical method. The ones we discuss in this chapter combine the two approaches into a single acquisition function.

4.2.1 Constrained Expected Improvement

A natural extension of the unconstrained acquisition functions to the case of constrained optimization is to impose some sort of restriction on where the unconstrained acquisition function can search. A simple, and intuitive, way to achieve this goal is to take an unconstrained acquisition function and weight its function value by the probability that the input satisfies the constraint. Perhaps the most well known use of this idea is that of constrained expected improvement (Schonlau et al. 1998; Gardner et al. 2014). As the name suggests, the unconstrained component of the acquisition function relies upon the EI acquisition function of Sect. 3.3.2, and for a given input, its corresponding EI value is weighted by the probability that the input satisfies the constraints. Thus, the constrained expected improvement (CEI) acquisition function can be formulated as follows

$$a_{\text{CEI}}(x) = \mathbb{E}\{I(x)\} \times \Pr(c(x) \leq 0), \tag{4.2}$$

where $\Pr(c(x) \leq 0)$ is the probability of satisfying the constraint. Here, the improvement function, $I(x)$, uses an f_{min}^n defined over the region of the input space where the constraint functions are satisfied since we are only concerned with improving upon the current solution in the valid region of the input space.

Fortunately, the derivation of $\mathbb{E}\{I(x)\}$ in equation (3.9) still holds in the constrained case. And so, all we are left to deal with is how to handle modeling the probability of satisfying the constraint. There is no one universal model used for calculating the probability, but rather, the choice of modeling strategy is usually dependent upon the type of values the constraint function returns. The two type of constraint functions that exists are those that return a continuous value and those that return a binary value. In the case of a continuous constraint, the constraint function provides a real-valued measure of constraint satisfaction. When the input does not satisfy the constraint the value returned gives a sense of how far away the input is to satisfying the constraint. For a continuous constraint function, it is simple enough to take the same approach that we do for the objective function, and model the constraint function output using a Gaussian process. For a single constraint, one can use the posterior predictive distribution of the Gaussian process in order to calculate the $\Pr(c(x) \leq 0)$, i.e.,

$$\Pr(c(x) \leq 0) = \int_{-\infty}^{0} N(Y_c(x); \mu_c(x), \sigma_c(x)) dY_c(x)$$

$$= \Phi\left(\frac{-\mu_c(x)}{\sigma_c(x)}\right) \tag{4.3}$$

where $Y_c(x)$ is the Gaussian process surrogate model for the constraint function, and this integral reduces to the Gaussian cumulative distribution function $\Phi(\cdot)$. In the case of multiple constraints, if we assume that the constraints are conditionally independent given x, then the probability of satisfying the constraints factorizes into

$$\Pr(c_1(x) \leq 0, \ldots, c_m(x) \leq 0) = \prod_{i=1}^{m} \Pr(c_i(x) \leq 0), \tag{4.4}$$

and everything proceeds as before in the case of the single constraint. In the case of dependent constraints, the joint probability $\Pr(c_1(x) \leq 0, \ldots, c_m(x) \leq 0)$ can be computed using numerical methods (Cunningham et al. 2013).

On the other hand, a continuous Gaussian process does little good when the constraint function only returns a binary outcome which specifies whether or not the constraint was satisfied. In the case of binary constraints, the available modeling choices become somewhat more interesting in that any classification model that can calculate the probability of belonging to one of two classes can be used. Here, in keeping with the spirit of BO, a natural choice might be to use a Gaussian process classification model, although other popular choices include such things as tree based classifiers like random forests (Breiman 2001) or Bayesian additive regression trees (BART) (Chipman et al. 2010), to name a few. Readers interested in learning more about Gaussian process classification models are encouraged to see (Rasmussen and Williams 2006).

Example Consider the following two-dimensional constrained optimization problem

$$\min \ f(x_1, x_2) = 4x_1^2 - x_1 - x_2 - 2.5$$

$$\text{s.t. } c_1(x_1, x_2) = -x_2^2 + 1.5x_1^2 - 2x_1 + 1$$

$$c_1(x_1, x_2) = 3x_1^4 + x_2^2 - 2x_1 - 4.25$$

where $-1.5 \leq x_1 \leq 2.5$, and $-3 \leq x_2 \leq 3$. The optimal solution to the constrained problem is $f(x_1, x_2) = -4.6958$, which occurs along the border of the valid region at $(x_1, x_2) = (0.1708, 2.1417)$ (see Fig. 4.1). Although the functional forms are known in this example, we will treat the objective function, $f(x)$, as if it were an

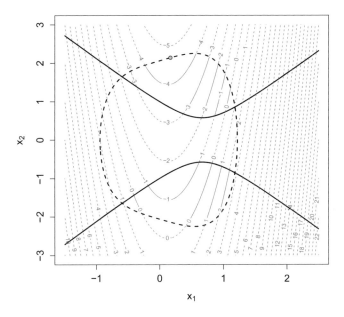

Fig. 4.1 Contours of the objective function colored by the two constraints. The solid black line denotes one constraint function, while the dashed black line denotes the other constraint function. Contours that are red are areas where the constraints are not satisfied, while green contours indicate areas where the constraints are satisfied. The blue point represents the global solution to the problem

expensive black-box function, and similarly for the two constraint functions, $c_1(x)$ and $c_2(x)$ so that we may solve it using constrained BO.

Following the general constrained BO algorithm outlined in Algorithm 2 of Sect. 4.1, we start with an initial LHS sample of size $n = 10$, and sequentially select 50 more inputs to evaluate based on the CEI acquisition function. Here we choose to model the objective and constraint functions using independent Gaussian process surrogate models, i.e., $Y_f(x)$ for the objective function, and $Y_{c_1}(x)$ and $Y_{c_2}(x)$ for the constraint functions. This modeling choice allows us to pick the next input to evaluate by maximizing the following easy to evaluate CEI acquisition function

$$a_{CEI}(x) = \mathbb{E}\{I(x)\} \times \Pr(c(x) \leq 0)$$

$$= \left[(f_{min}^n - \mu_f(x))\Phi\left(\frac{f_{min}^n - \mu_f(x)}{\sigma_f(x)}\right) + \sigma_n(x)\phi\left(\frac{f_{min}^n - \mu_f(x)}{\sigma_f(x)}\right) \right]$$

$$\times \left[\prod_{i=1}^{2} \Phi\left(-\frac{\mu_{c_i}(x)}{\sigma_{c_i}(x)}\right) \right]. \tag{4.5}$$

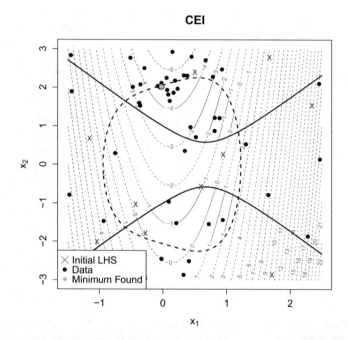

Fig. 4.2 A view of the performance of the BO algorithm using the constrained expected improvement (CEI) acquisition function for a single run of a Monte Carlo experiment

Given the setup, we proceed to search for the global solution to constrained optimization problem using the CEI acquisition function to choose the next 50 inputs to sequentially evaluate. Figure 4.2 shows the performance of the BO algorithm using the constrained expected improvement (CEI) acquisition function in order to give an idea of how the CEI acquisition searches the input space. Constrained optimization in this problem is hard due to the fact that the valid regions are small relative to the input space and disconnected. We see that with a starting LHS of size $n = 10$, only two inputs are selected in the valid region and that of the two disconnected valid regions, only one of them has any initial data. However, even without starting knowledge of the second valid region, the CEI algorithm performs quite well spending a lot of its effort searching near the global solution. Note that the CEI acquisition is just expected improvement weighted by the probability of being in the valid region, and so the CEI acquisition function can tend to explore invalid regions often when the estimate of the probability is either poor, or simply because the expected improvement is very high in the invalid region.

Obviously, the performance of the acquisition function for guiding the BO algorithm is highly dependent upon the initial sample used to initialize the algorithm. To get an idea of the performance, as well as the robustness, of the BO algorithm under the CEI acquisition function, we rerun the BO algorithm within 30 Monte Carlo experiments. The results of the 30 Monte Carlo experiments is captured in Fig. 4.3.

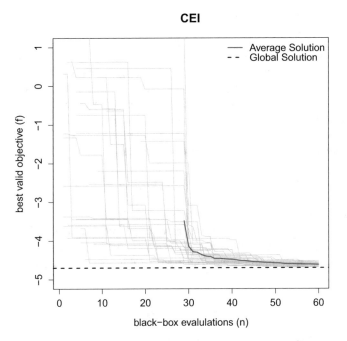

Fig. 4.3 A view of the performance of the BO algorithm using the constrained expected improvement (CEI) acquisition function for the 30 the Monte Carlo experiments. Here, each grey line represents the best value found over the search by the BO algorithm during a single run of the Monte Carlo experiment. The red average line starts when all of the 30 Monte Carlo experiments have found a valid solution

Note that, unlike the Monte Carlo progress plots in Chap. 3, the best valid average solution over the 30 Monte Carlo experiments need not be calculable for all of the black-box evaluations. This is due to the fact that each Monte Carlo experiment will potentially find its first valid objective function evaluation at a different point in time in the sequential evaluation process. For example, it is clear to see that the individual trajectories of the best valid solution (i.e., grey lines) all do not start at the same time and point. In fact, at least one of the 30 Monte Carlo initial LHSs is composed entirely of inputs selected in the invalid region, and so clearly the BO algorithm may even start with no knowledge of the valid regions. As we can see from Fig. 4.3, the worst of the 30 Monte Carlo experiments does not begin selecting inputs from the valid region until about the thirtieth input evaluation (the start of the red average solution curve). Thus, being able to locate or hone in on the valid region of the input space is critical to the success of the BO algorithm. We note that many publications in the literature assume knowledge of at least one point in the valid region, for example, by discarding an initial LHS without any valid points and generating new ones until a sample is obtained with at least one valid point. Such an approach leads to prettier plots, but it hides the computational expense of generating the additional LHSs.

4.2.2 Asymmetric Entropy

As discussed in Sect. 4.2, sometimes it is important to try to keep as many runs as possible inside the valid region while searching for the optimum. At the same time, the minimum may be expected to be along the constraint boundary, because the constraints are operating in opposition to the objective function. A well-studied approach for focusing on a boundary is to use entropy as a utility function.

Consider the problem of finding the constraint boundary by estimating the probability an input x will be inside the valid region, i.e.,

$$p(x) = \Pr(c(x) \le 0). \tag{4.6}$$

We can estimate the boundary by finding x such that $p(x) = 0.5$. It turns out that we can recast the boundary-finding problem as an unconstrained BO problem. We can create a surrogate model for p, such as a classification Gaussian process or any other classifier. A common utility function for this search would be the Shannon entropy:

$$S(x) = -p(x) \times \log(p(x)) - (1 - p(x)) \times \log(1 - p(x)). \tag{4.7}$$

Using the expected value of $S(x)$ under the surrogate as the acquisition function leads to a BO algorithm for finding the boundary.

Our actual problem of interest is constrained optimization, which here is focused on finding an optimum along a boundary. Thus we may want to include entropy as part of an acquisition function, such as taking a product of EI and entropy. We might want to more heavily weight either EI or entropy, and thus a family of acquisition functions is

$$a(x) = EI(x)^{\omega_1} \times S(x)^{\omega_2}, \tag{4.8}$$

where ω_1 and ω_1 act as weights for EI and entropy, respectively. As this acquisition function combines EI and entropy, it will tend to explore the promising space just beyond the boundary, primarily in the invalid region, where EI will be larger. The entropy term pulls it closer to the boundary, but not enough to pull it into the valid region. To address this imbalance, Lindberg and Lee (2015) introduced an acquisition function based on asymmetric entropy, which can focus the search inside the valid region. Asymmetric entropy was originally created by Marcellin et al. (2006) in the context of fitting decision trees when one class is relatively uncommon and thus can benefit from being favored in the tree growth algorithm. Asymmetric entropy is given by

$$S_a(x) = \frac{2p(x)(1 - p(x))}{p(x) - 2wp(x) + w^2}, \tag{4.9}$$

where w is a mode location parameter. Maximum asymmetric entropy is achieved at $p = w$ instead of the usual $p = 0.5$. Thus selecting $w > 0.5$ will push exploration more toward the valid region. Lindberg and Lee (2015) recommend $w = 2/3$. Substituting asymmetric entropy into the acquisition function, we now have

$$a_{AE}(x) = EI(x)^{\omega_1} \times S_a(x)^{\omega_2}.$$

Lindberg and Lee (2015) recommend $\omega_1 = 1$ and $\omega_2 = 5$, giving the acquisition function

$$a_{AE}(x) = EI(x) \times S_a(x)^5. \tag{4.10}$$

This formulation can be effective at solving the constrained optimization problem while searching points primarily inside the valid region.

Example Returning to the two-dimensional constrained optimization example in Sect. 4.2.1, consider solving again the following problem

$$\min \ f(x_1, x_2) = 4x_1^2 - x_1 - x_2 - 2.5$$

$$\text{s.t. } c_1(x_1, x_2) = -x_2^2 + 1.5x_1^2 - 2x_1 + 1$$

$$c_1(x_1, x_2) = 3x_1^4 + x_2^2 - 2x_1 - 4.25$$

where $-1.5 \leq x_1 \leq 2.5$, and $-3 \leq x_2 \leq 3$. Proceeding as before, we start with an initial LHS sample of size $n = 10$, and sequentially select 50 more inputs to evaluate based on the AE acquisition function. Given that both constraint functions return a continuous value, we choose to model the constraints using independent Gaussian process surrogate models, $Y_{c_1}(x)$ and $Y_{c_2}(x)$, and their respective predictive means and variances to calculate the probability $p(x)$ as

$$p(x) = \Pr(c(x) \leq 0) = \Phi\left(-\frac{\mu_{c_1}(x)}{\sigma_{c_1}(x)}\right) \times \Phi\left(-\frac{\mu_{c_2}(x)}{\sigma_{c_2}(x)}\right). \tag{4.11}$$

We then can calculate the asymmetric entropy by plugging $p(x)$ into $S_a(x)$ and setting $w = 2/3$. Likewise, we follow the recommendation set forth in Lindberg and Lee (2015) and set $\omega_1 = 1$, and $\omega_2 = 5$ to finish the specification of the acquisition function.

Running the BO algorithm under these settings, Fig. 4.4 shows the performance of the BO algorithm using the AE acquisition function. Recall that when $w > 0.5$, asymmetric entropy will push the search towards the inside of the boundary of the valid region. What we see in Fig. 4.4 is exactly that. When searching within the valid regions, the AE acquisition function pushes the search towards the boundaries of the valid regions when exploring.

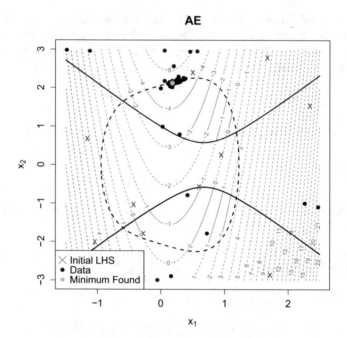

Fig. 4.4 A view of the performance of the BO algorithm using the asymmetric entropy (AE) acquisition function for a single run of a Monte Carlo experiment

To get an idea of the overall performance of the BO algorithm using the AE acquisition function, we run 30 Monte Carlo experiments of the BO algorithm. Figure 4.5 shows the individual and average performance of the BO algorithm over the 30 Monte Carlo experiments. It is not until after the twentieth input evaluation that the worst of the Monte Carlo experiments has found an input in a valid region; however, every single solution search path seems to hone in on the global solution very quickly. By about the twenty fifth iteration of the algorithm, all of the Monte Carlo runs have converged to the global minimum of the problem. Clearly, using the AE acquisition function is advantageous when the solution to the constrained optimization problem lies along the boundary of the input space since the natural behavior of the acquisition function is to want to search for the boundary and explore along it.

4.2.3 Augmented Lagrangian

Augmented Lagrangian methods (Bertsekas 1982; Nocedal and Wright 2006) are a class of methods based on using a penalty function to combine constraint satisfaction with objective function optimization into a new single scalar acquisition

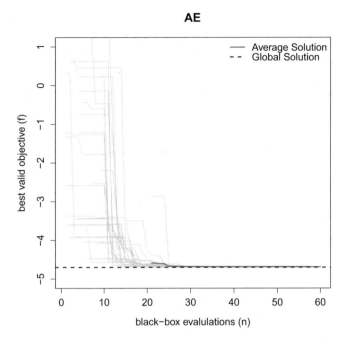

Fig. 4.5 A view of the performance of the BO algorithm using the asymmetric entropy (AE) acquisition function for the 30 the Monte Carlo experiments. Here, each grey line represents the best value found over the search by the BO algorithm during a single run of the Monte Carlo experiment. The red average line starts when all of the 30 Monte Carlo experiments have found a valid solution

function, which can then be solved as a sequence of unconstrained optimization problems. The Augmented Lagrangian approach combines the original objective function with a penalty parameter multiplied by the constraint function, plus a Lagrangian term, and seeks to minimize that combination. We work here with the negative of that term as the acquisition function that we maximize:

$$a_{\mathrm{AL}}(x; \lambda, \rho) = -f(x) - \lambda^T c(x) - \frac{1}{2\rho} \sum_{j=1}^{m} \max(0, c_j(x))^2, \qquad (4.12)$$

where f is the objective function, c is the vector of constraint functions, $\rho > 0$ is the penalty parameter, and $\lambda \in \mathbb{R}^m_+$ is the Lagrange multiplier. As we search the space, for given values of λ and ρ, we choose the next point x^* as

$$x^* = \underset{x \in \mathcal{X}}{\mathrm{argmax}} \, a_{\mathrm{AL}}(x; \lambda, \rho). \qquad (4.13)$$

If we were to fix λ and ρ, this approach would convert the constrained optimization problem into an unconstrained optimization problem, typically simplifying the problem. In practice, we use a sequence of values for λ and ρ, and so we have converted a constrained problem to a sequence of unconstrained problems.

The augmented Lagrangian approach has good theoretical convergence properties when $\rho \rightarrow 0$. However, as $\rho \rightarrow 0$, the problem becomes increasingly ill-conditioned, and difficult to solve numerically. Thus real-world algorithms use a sequence of values for ρ whose limit is 0. As ρ is updated, λ is updated at iteration k as a typical Lagrange multiplier

$$\lambda_j^k = \max\left(0, \lambda_j^{k-1} + \frac{1}{\rho^{k-1}} c_j(x^k)\right). \tag{4.14}$$

Augmented Lagrangian methods were developed as numerical methods, designed for direct numerical optimization. Gramacy et al. (2016) adapted the augmented Lagrangian approach for Bayesian optimization by incorporating Gaussian process surrogate modeling into the algorithm. An independent Gaussian process is used to approximate f and each constraint c_j, for a total of $m + 1$ Gaussian process models. In particular, Y_f is a Gaussian process surrogate for f and Y_{c_j} is a Gaussian process surrogate for c_j. These surrogates are important because (4.13) requires the selection of the next point, x^*, based on the unknown f and c. The Gaussian process surrogates are used to guide this choice of x^*. The acquisition function is thus approximated with:

$$-Y_f(x) - \lambda^T Y_c(x) - \frac{1}{2\rho} \sum_{j=1}^m \max(0, Y_{c_j}(x))^2. \tag{4.15}$$

Given this approximation, how does one choose the next x^* in a BO routine? As we update our Gaussian process surrogate, Y will have a distribution derived from the posterior distributions of each of the Gaussian processes, and so Y is not a scalar function that can be directly minimized. Gramacy et al. (2016) suggest several methods for guiding this choice, stemming from two conceptual approaches: following the predictive mean, or expected improvement.

The first approach for choosing the point, x^*, that maximizes (4.12) based on the approximation (4.15) is to choose the x^* that maximizes the posterior predictive mean of (4.15). At any point x, $Y_f(x)$ will have a posterior predictive mean that is Gaussian with mean $\mu_f(x)$ and variance $\sigma_f^2(x)$. Similarly, each $Y_{c_j}(x)$ will have a Gaussian posterior predictive mean with mean $\mu_{c_j}(X)$ and variance $\sigma_{c_j}^2(x)$. So we can now write

$$\mathbb{E}\left[a_{\text{AL}}(x)\right] \approx -\mathbb{E}\left[Y_f(x)\right] - \lambda^T \mathbb{E}\left[Y_c(x)\right] - \frac{1}{2\rho} \sum_{j=1}^m \mathbb{E}\left[\max(0, Y_{c_j}(x))^2\right]$$

$$= -\mu_f(x) - \lambda^T \mu_c(x) - \frac{1}{2\rho} \sum_{j=1}^m \mathbb{E}\left[\max(0, Y_{c_j}(x))^2\right]. \tag{4.16}$$

Gramacy et al. (2016) provide an expansion of that last expectation as

$$\mathbb{E}\left[\max(0, Y_{c_j}(x))^2\right] = \sigma_{c_j}^2(x)\left\{\left(1+\left(\frac{\mu_{c_j}(x)}{\sigma_{c_j}(x)}\right)^2\right)\Phi\left(\frac{\mu_{c_j}(x)}{\sigma_{c_j}(x)}\right)+\phi\left(\frac{\mu_{c_j}(x)}{\sigma_{c_j}(x)}\right)\right\},$$

$$\text{(4.17)}$$

where Φ and ϕ are the cumulative distribution function and probability density function for the standard Gaussian distribution, respectively.

The second approach is based on expected improvement, choosing the x^* that has the largest expected improvement in Y. Because this EI is not available in closed form, Gramacy et al. (2016) suggest a Monte Carlo approximation. Draw T Monte Carlo samples $y_f^{(t)}(x), y_{c_1}^{(t)}(x), \ldots, y_{c_m}^{(t)}(x)$ from Gaussian distributions $N\left(\mu_f(x), \sigma_f^2(x)\right)$ and $N\left(\mu_{c_j}(x), \sigma_{c_j}^2(x)\right)$, for $t = 1, \ldots, T$. The approximation is

$$\mathbb{E}\left[\mathbb{1}_Y(x)\right] \approx \frac{1}{T}\sum_{t=1}^{T}\max\left\{0, y_{\min}-\left[y_f^{(t)}(x)+\lambda^T y_c^{(t)}(x)+\frac{1}{2\rho}\sum_{j=1}^{m}\max\left(0, y_{c_j}^{(t)}(x)\right)^2\right]\right\}$$

$$\text{(4.18)}$$

where $y_{\min}$ is the smallest value of (4.15) that has been observed across all previous iterations.

For both approaches, full optimization would be impractical, and x^* can be chosen by drawing a random sample of candidates and choosing the x^* which is the best among the candidates evaluated. Gramacy et al. (2016) provide additional discussion on generating improved candidate sets. They also discuss variations on these two approaches that approximate the acquisition functions by removing the max operator, which leads to simplified expressions that can be written in closed form in certain cases.

Example Using the AL acquisition function, let us again solve the following two-dimensional constrained optimization problem

$$\min\ f(x_1, x_2) = 4x_1^2 - x_1 - x_2 - 2.5$$

$$\text{s.t.}\ c_1(x_1, x_2) = -x_2^2 + 1.5x_1^2 - 2x_1 + 1$$

$$c_1(x_1, x_2) = 3x_1^4 + x_2^2 - 2x_1 - 4.25$$

where $-1.5 \leq x_1 \leq 2.5$, and $-3 \leq x_2 \leq 3$. Starting with an initial LHS of size $n = 10$, we run the BO algorithm using the AL acquisition function to sequentially choose the next 50 inputs to evaluate. Figure 4.6 shows the performance of the BO algorithm over a single Monte Carlo run. The AL acquisition function performs quite well in this example, spending most of its time exploring input space of one of the valid regions. Unlike the AE acquisition function, the AL acquisition function

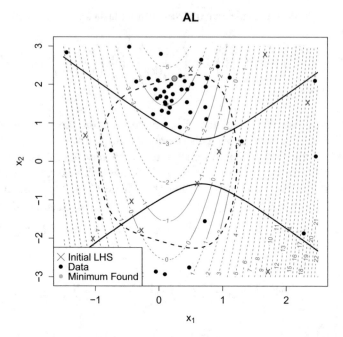

Fig. 4.6 A view of the performance of the BO algorithm using the augmented Lagrangian (AL) acquisition function for a single run of a Monte Carlo experiment

does not have a preference for solely trying to search the boundary of the input space and so it takes a more exploratory approach to searching the valid region.

Similar to the CEI acquisition function, it takes about 30 input evaluations before all of the 30 Monte Carlo experiments find a valid input (Fig. 4.7). However, the AL acquisition function still does a quite good job at converging, on average, to the global solution of the problem.

4.2.4 Barrier Methods

Barrier methods (Nocedal and Wright 2006), also known as interior point methods, are a natural strategy for solving black-box constrained optimization problems as they try to decrease the objective function as much as possible while ensuring that the boundary of the constraint space is never crossed. In order to ensure that the boundary of the constraint space is never crossed, barrier methods replace the inequality constraints in the constrained optimization problem in (4.1) with an extra term in the objective function that can be viewed as a penalty for approaching the

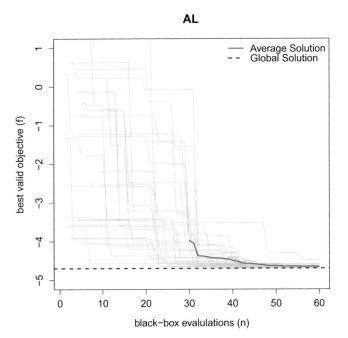

Fig. 4.7 A view of the performance of the BO algorithm using the augmented Lagrangian (AL) acquisition function for the 30 the Monte Carlo experiments. Here, each grey line represents the best value found over the search by the BO algorithm during a single run of the Monte Carlo experiment. The red average line starts when all of the 30 Monte Carlo experiments have found a valid solution

boundary. Here, the constrained optimization problem in (4.1) can be re-written as the following unconstrained optimization problem

$$\min_{x} \left\{ f(x) + \sum_{i=1}^{m} \mathbf{B}_{\{c_i(x) \leq 0\}}(x) \right\}, \tag{4.19}$$

where $\mathbf{B}_{\{c_i(x) \leq 0\}}(x) = 0$ if $c_i(x) \leq 0$ and ∞ otherwise. Although mathematically equivalent, the introduction of $\mathbf{B}_{\{c_i(x) \leq 0\}}(x)$ in the reformulation of the original constrained optimization problem is not particularly useful as it introduces an abrupt discontinuity when $c_i(x) > 0$. This discontinuity eliminates the use of calculus to minimize (4.19). To remedy this issue, the discontinuous function in (4.19) can be replaced with a continuous approximation, $\xi(x)$, that is ∞ when $c_i(x) > 0$ but is finite for $c_i(x) \leq 0$. This continuous approximation, $\xi(x)$, is referred to as the barrier function as it will create a "barrier" to exiting the valid region for the

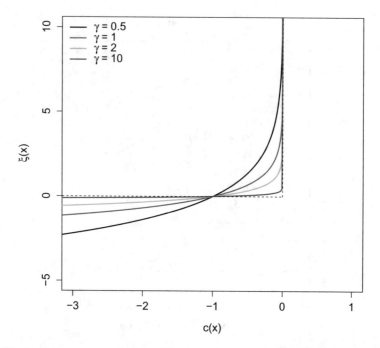

Fig. 4.8 As γ approaches ∞, the log barrier function, $\xi(x)$, becomes a better approximation to $\mathbf{B}_{\{c_i(x)\leq 0\}}(x)$ (i.e., the dashed line)

search algorithm. Many barrier functions exists, however, a popular choice of barrier function is the log barrier function which is defined as

$$\xi(x) = -\left(\frac{1}{\gamma}\right) \sum_{i=1}^{m} \log(-c_i(x)) \qquad (4.20)$$

for $\gamma > 0$. Note that the log barrier function, $\xi(x)$, is a smooth approximation of $\sum_{i=1}^{m} \mathbf{B}_{\{c_i(x)\leq 0\}}(x)$ when $c_i(x) < 0$, and that this approximation improves as γ goes to ∞ (see Fig. 4.8). Now, replacing $\mathbf{B}_{\{c_i(x)\leq 0\}}(x)$ with the log barrier function, $\xi(x)$, we can approximate the problem in (4.19) as

$$\min_{x}\{B(x;\gamma)\} = \min_{x}\left\{f(x) - \left(\frac{1}{\gamma}\right) \sum_{i=1}^{m} \log(-c_i(x))\right\}. \qquad (4.21)$$

Solving the minimization problem in (4.21) becomes a much more manageable and tractable problem as compared to the minimization problem in (4.19).

Recognizing the attractive qualities of barrier methods for constrained optimization, Pourmohamad and Lee (2021) extended barrier methods to the BO framework by modeling the quantity $B(x;\gamma)$ in (4.21), using independent Gaussian process surrogates $Y_f(x)$ and $Y_c(x) = (Y_{c_1}(x), \ldots, Y_{c_m}(x))$ for the objective and constraint functions, i.e.

$$Y(x) = Y_f(x) - \left(\frac{1}{\gamma}\right) \sum_{i=1}^{m} \log(-Y_{c_i}(x)). \tag{4.22}$$

Pourmohamad and Lee (2021) suggests that optimization can then proceed by minimizing the predictive mean surface of $Y(x)$. At first glance, this may sound like a poor idea since minimizing the expectation of a Gaussian process typically leads to a greedy search algorithm (i.e., think back to the discussion in Sect. 3.2 about minimizing the predictive mean of the Gaussian process, $\mu_n(x)$). However, as will be shown, in this case minimizing the predictive mean surface of $Y(x)$ leads to an acquisition function that searches the space both locally and globally. The result of minimizing the predictive mean surface of $Y(x)$ results in:

$$\min_{x} \mathbb{E}(Y(x)) \approx \min_{x} \mu_f(x) - \left(\frac{1}{\gamma}\right) \sum_{i=1}^{m} \left(\log(-\mu_{c_i}(x)) + \frac{\sigma_{c_i}^2(x)}{2\mu_{c_i}^2(x)} \right) \tag{4.23}$$

The details of the derivation of the expectation in (4.23) can be found in Pourmohamad and Lee (2021). To finally recast this in the language of BO, we instead pivot to maximizing the negative value of this equation and thus establish the barrier method (BM) acquisition function, i.e.,

$$a_{\mathrm{BM}}(x) = -\mu_f(x) + \left(\frac{1}{\gamma}\right) \sum_{i=1}^{m} \left(\log(-\mu_{c_i}(x)) + \frac{\sigma_{c_i}^2(x)}{2\mu_{c_i}^2(x)} \right). \tag{4.24}$$

Two problems arise from this acquisition function. The first problem is that there is no explicit rule, in the context of BO, on how to set γ. In the mathematical programming literature (e.g., Nocedal and Wright (2006)), it is common practice to have the value of $\gamma \to \infty$ such that, at iteration $k + 1$ of the barrier method, $\gamma_{k+1} > \gamma_k$. In effect, this leads to steadily decreasing the penalty for approaching the boundary of the valid region throughout the optimization. However, much like the LCB acquisition function of Sect. 3.3.3, γ is still a tuning parameter left to user's discretion. The second problem is that (4.24) contains no variability term associated with the objective function, but rather only with the constraints, i.e., $\sigma_{c_i}^2(x)$. Without a term like $\sigma_f^2(x)$ in (4.24) to measure the prediction uncertainty for the objective function, the acquisition function will tend to favor exploitation, rather than exploration, as it will assume that it is predicting the objective function at untried inputs without error. Solving both of these problems at once, Pourmohamad and Lee (2021) recommended setting $\gamma = 1/\sigma_f^2(x)$, where $\sigma_f^2(x)$ is the predictive variance associated with the Gaussian process surrogate model for the objective function f. This leads to the revised acquisition function

$$a_{\mathrm{BM}}(x) = -\mu_f(x) + \sigma_f^2(x) \sum_{i=1}^{m} \left(\log(-\mu_{c_i}(x)) + \frac{\sigma_{c_i}^2(x)}{2\mu_{c_i}^2(x)} \right), \quad (4.25)$$

which (1) gives a rule for setting γ based on the current level of uncertainty in the predictions, and (2) injects an uncertainty term for the objective function into the acquisition function.

A second approach was also proposed in Pourmohamad and Lee (2021) which was to replace the surrogate model for the objective function, $Y_f(x)$, with the improvement function $-I(x)$ in (4.22), i.e.,

$$\min_{x} \mathbb{E}\left(-I(x) - \left(\frac{1}{\gamma}\right) \sum_{i=1}^{m} \log(-c_i(x)) \right) = \min_{x} -\mathbb{E}(I(x)) - \left(\frac{1}{\gamma}\right) \sum_{i=1}^{m} \left(\log(-\mu_{c_i}) + \frac{\sigma_{c_i}^2}{2\mu_{c_i}^2} \right).$$
$$(4.26)$$

The idea here being that if you instead take the expectation of the improvement function in (4.26), that you would incur all of the benefits of the expected improvement acquisition function, i.e., a variance term for the objective function and the natural exploration-exploitation search characteristics. Note that since we are minimizing in (4.23) we will need to use the negative improvement function. The minimization problem in (4.26) leads to the following acquisition function

$$a_{\mathrm{BM}}(x) = (f_{\min}^n - \mu_f(x)) \Phi\left(\frac{f_{\min}^n - \mu_f(x)}{\sigma_f(x)} \right) + \sigma_f(x) \phi\left(\frac{f_{\min}^n - \mu_f(x)}{\sigma_f(x)} \right)$$
$$+ \left(\frac{1}{\gamma}\right) \sum_{i=1}^{m} \left(\log(-\mu_{c_i}) + \frac{\sigma_{c_i}^2(x)}{2\mu_{c_i}^2(x)} \right), \quad (4.27)$$

where again, Pourmohamad and Lee (2021) suggest setting $\gamma = 1/\sigma_f^2(x)$.

Example Using the BM acquisition function, we solve one last time the following two-dimensional constrained optimization problem

$$\min \ f(x_1, x_2) = 4x_1^2 - x_1 - x_2 - 2.5$$
$$\text{s.t. } c_1(x_1, x_2) = -x_2^2 + 1.5x_1^2 - 2x_1 + 1$$
$$c_1(x_1, x_2) = 3x_1^4 + x_2^2 - 2x_1 - 4.25$$

where $-1.5 \le x_1 \le 2.5$, and $-3 \le x_2 \le 3$. Once again, we start with an initial LHS of size $n = 10$, and sequentially pick 50 additional inputs to evaluate based on the BM acquisition function. For illustration here, we shall use the form of the acquisition function in (4.25). The performance of the BM acquisition function

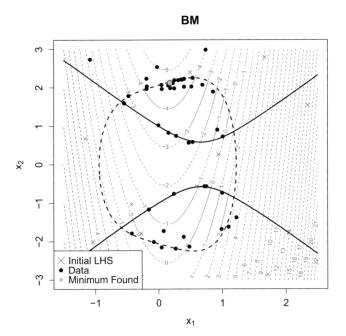

Fig. 4.9 A view of the performance of the BO algorithm using the barrier method (BM) acquisition function for a single run of a Monte Carlo experiment

for guiding the BO algorithm can be seen in Fig. 4.9. As opposed to the CEI, AE, and AL acquisition functions, the BM acquisition places higher importance on trying to stay within the valid region, and so exploring the invalid regions far less. This characteristic can be both desirable and undesirable. Staying within the valid region makes a lot of sense since we are concerned with finding a valid solution to the problem and, based on how computationally expensive the computer model is, searching in the invalid region can be viewed as wasteful since those inputs will not be the solution to the optimization problem and their evaluation is costly. On the other hand, as seen in Sects. 4.2.2 and 4.2.3, evaluating inputs in the invalid region is beneficial with helping the Gaussian process surrogate models learn both the objective and constraint function surfaces better, which ultimately leads to better prediction and uncertainty reduction, both of which are important components of a good acquisition function. Much like the AE acquisition function, the BM acquisition as well has a tendency to explore the boundary of the valid regions. Note though that while the original barrier methods were designed to never cross the border of the valid region, the BM acquisition function explores along the boundary, sometimes crossing it, since the location of the boundary is being estimated, and the Gaussian process surrogate model learns the boundary by sometimes going just beyond it.

AL

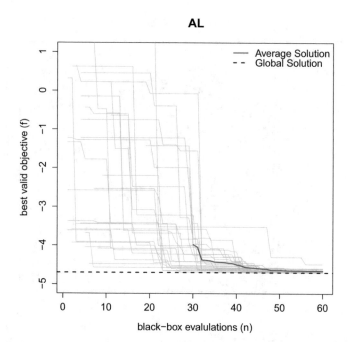

Fig. 4.10 A view of the performance of the BO algorithm using the barrier method (BM) acquisition function for the 30 the Monte Carlo experiments. Here, each grey line represents the best value found over the search by the BO algorithm during a single run of the Monte Carlo experiment. The red average line starts when all of the 30 Monte Carlo experiments have found a valid solution

Figure 4.10 captures the behavior of 30 Monte Carlo experiments for running the BO algorithm using the BM acquisition function. Overall, the performance of the BM acquisition function is quite good, with all of the 30 different Monte Carlo runs obtaining a valid input by about the twentieth input evaluation. After the initial LHS, the BM acquisition steadily guides the BO algorithm to the global solution of the problem on average.

For sake of comparison, we plot the average performance of the CEI, AE, AL, and BM algorithms over their respective 30 Monte Carlo experiments (Fig. 4.11). Here, the initial 30 LHSs, across the Monte Carlo experiments, are the same starting inputs for each acquisition function. What we see from Fig. 4.11 is vastly different average performance of the BO algorithm under the four different acquisition functions. Here, BM and AE are much better at finding valid inputs earlier on which is due to their tendencies to approach, or stay within, the boundaries of the valid region. CEI and AL are slower to find valid inputs due to the fact that they will allow for more exploration of the invalid region as compared to the AE and BM acquisition functions, i.e., there is no heavy penalty for exiting the valid region or need to search out the boundary of the valid region. Although the CEI and AL acquisition functions are slower to find valid starting inputs, they still perform as

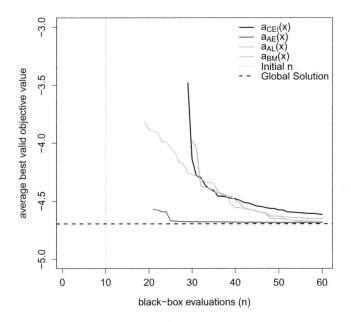

Fig. 4.11 The average performance of the CEI, AE, AL, and BM acquisition functions over the 30 Monte Carlo experiments

well as the BM acquisition function in this example. On the other hand, the AE acquisition function clearly dominates the other three acquisition functions with regards to the number of function evaluations needed to converge to the global solution of the problem. This superior performance is due in part to the fact that the global solution lies exactly on the border of the valid region which is exactly where the AE acquisition function wants to search. We emphasize that this a single example, and that on other examples, a different one of these methods may perform best.

4.3   Constrained Sprinkler Computer Model

Recall that in Sect. 3.4 we maximized the range of the garden sprinkler via an unconstrained optimization problem. Given that the garden sprinkler computer model returns a multi-objective output, we can now optimize one of the objective function outputs subject to the constraint that either one or both of the other objective function values is above (or below) a certain value. For sake of example, let us assume that still want to maximize the range of the garden sprinkler but now subject to the constraint that the water consumption must not be greater than five. Here we will assume that constraining the speed of the garden sprinkler (i.e., the third output of the garden sprinkler computer model) is not of concern. Casting the garden sprinkler computer model in the framework of a constrained optimization, we formulate the problem as follows:

$$x^* = \underset{x \in \mathcal{X}}{\operatorname{argmin}}\{-f(x)\} \text{ subject to } c(x) - 5 \leq 0, \tag{4.28}$$

Here the objective function, $f(x)$, describes the range at which the garden sprinkler can spray water, while the constraint function, $c(x)$, determines whether an acceptable amount of water is used. Note that since we wish to maximize the range of the garden sprinkler, we shall instead minimize the negative objective function in order to find the input that maximizes it. The inputs $x = (x_1, \ldots, x_8)^T \in \mathcal{X}$ represent the eight physical attributes of the garden sprinkler (see Fig. 1.6 and Table 1.1) that can be set within the computer model. The computer model is essentially a black-box function since, for any input configuration evaluated by the model, the only information that is returned is that of the objective and constraint values.

Now, we shall solve for the maximum value of the range of the garden sprinkler, subject to the water consumption constraint, using the constrained expected improvement (CEI), asymmetric entropy (AE), augmented Lagrangian (AL), and barrier method (BM) acquisition functions, and shall compare and contrast their performances. We will initialize the BO algorithm using a LHS of size $n = 10$, and sequentially evaluate an addition 90 inputs for a total computational budget of 100 input evaluations. In order to assess the robustness of the solutions of the BO algorithm, under the four different acquisition functions, we repeat solving this constrained optimization problem using 30 Monte Carlo experiments. Figure 4.12 shows the results of the 30 Monte Carlo experiments for a given acquisition function.

Visually, it looks like the CEI, AL, and BM acquisition functions all have similar performance, while the AE acquisition perhaps has a few better runs of the Monte Carlo experiments (i.e., lower best valid objective values) as well as fewer worst solutions. For each acquisition function, taking the average of the solutions over the 30 Monte Carlo experiments reveals that the AE acquisition function indeed performed better than the other acquisition functions (Fig. 4.13). Although it took much longer, on average, for the AE acquisition to start evaluating valid inputs, it still was capable of finding a significantly better solution than the other three acquisition functions. On the other hand, the average performance of the CEI, AL, and BM acquisition functions was very similar with perhaps the exception of the BM acquisition function doing slightly better (i.e., lower average values) through several stretches of black-box iterations.

Table 4.1 gives a numerical summary of the performance of the four acquisition functions. Besides being better on average, the AE acquisition function also did the best with regards to both the best and worst final solutions found over the 30 Monte Carlo experiments as compared to the CEI, AL, and BM acquisition functions. The observed differences in worst and final solutions between the CEI, AL, and BM acquisition functions, over the 30 Monte Carlo experiments, were negligible, reaffirming the overall comparable performance of the three acquisition functions.

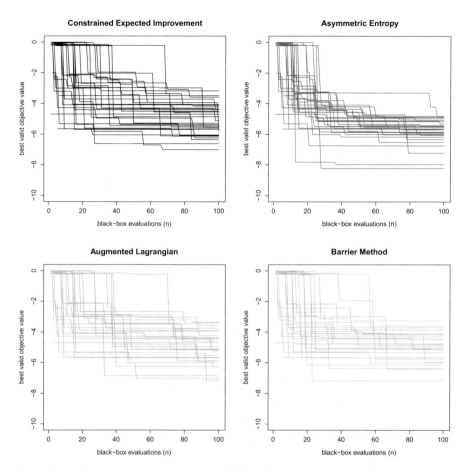

Fig. 4.12 A view of the performance of the BO algorithm, using the four different acquisition functions, for the 30 the Monte Carlo experiments. Here, each line represents the best value found over the search by the BO algorithm during a single run of the Monte Carlo experiment

Lastly, evaluating a LHS of size $n = 1,000,000$ inputs results in a global solution of -9.46. In this case, it is clear that the four acquisition functions have not yet converged to the global solution of the problem. This is consistent with the fact that in Fig. 4.13, all of the progress lines are still trending downward. Practically speaking, this means that the total budget for input evaluations needs to be increased past 100, and that the constrained BO algorithm may benefit from increasing the size of the initial LHS.

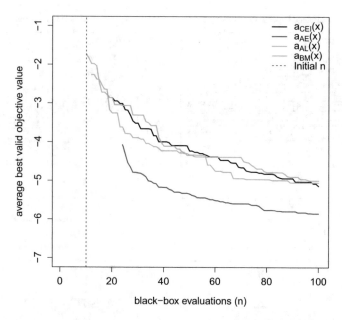

Fig. 4.13 The results of running 30 Monte Carlo experiments for each acquisition function. The plot shows the average best objective function values found over 100 black-box iterations

Table 4.1 The average, best, and worst solution found at the end of the 30 Monte Carlo experiments by each acquisition function

Acquisition function	Average final solution	Best final solution	Worst final solution
Constrained expected improvement	−5.14	−7.01	−3.17
Asymmetric entropy	−5.85	−8.22	−4.82
Augmented lagrangian	−5.00	−7.17	−3.36
Barrier method	−5.05	−7.16	−3.25

Chapter 5
Conclusions

This short book provided an introduction to the topic of Bayesian Optimization, including several examples of acquisition functions for each of unconstrained and constrained optimization. We emphasize that these are just a few choices among the many that have been created. And this is a vibrant field, with more new work continually emerging.

For both unconstrained and constrained optimization, we have selected some examples of acquisition functions that we have found to work well, and we also present one or two comparators that are less effective. For unconstrained optimization, random search is a baseline that all algorithms should be able to significantly outperform. Probability of Improvement is a reasonable concept, but it is not as effective as other more sophisticated methods. Both the Expected Improvement and the Lower Confidence Bound have been found to perform well. Neither method routinely outperforms the other; one can find different examples where each of them performs better.

Analogously for constrained optimization, Constrained Expected Improvement is a solid approach and a good baseline, but it is generally less effective than the other three methods. Among Asymmetric Entropy, Augmented Lagrangian, and Barrier Methods, none routinely outperform each other. While Asymmetric Entropy happened to have the best results in the two examples presented here, one of the other two methods might prevail on other examples.

Bayesian optimization is a very fast moving field within the machine learning community, with lots of evolving topics. Although not presented in this book, it would be remiss not to mention some of the current hot topics in Bayesian optimization. As of the writing of the book, two of the hottest growing topics in Bayesian optimization are batch BO and BO for high input dimensions. All of the work in this book revolved around BO approaches that explored the input space by sequential evaluation. On the other hand, batch BO recommends proposing batches of input values to explore simultaneously. Batch BO is especially advantageous when parallel computing can be employed. Readers interested in learning more

T. Pourmohamad, H. K. H. Lee, *Bayesian Optimization with Application to Computer Experiments*, SpringerBriefs in Statistics, https://doi.org/10.1007/978-3-030-82458-7_5

about batch BO are recommended to read the following works: Gonzalez et al. (2016), Hernández-Lobato et al. (2017), Kandasamy et al. (2018), Wang et al. (2018), Eriksson et al. (2019), and Mathesen et al. (2021), for example. With regards to BO for high input dimensions, the BO methods presented in this book have typically been applied to input dimensions, usually at most, of around 15 within the BO literature. BO for high input dimensions seeks to extend BO algorithms to the much higher dimensional settings, and in most cases, where the input evaluation budget is still relatively small. A lot of recent work has been done in this direction, and interested readers are encouraged to see the works of Binois et al. (2015), Wang et al. (2016), Eriksson et al. (2018), Mutny and Krause (2018), Oh et al. (2018), Rolland et al. (2018), Kirschner et al. (2019), Nayebi et al. (2019), Binois et al. (2020), and Letham et al. (2020), to a name a few.

Appendix A
R Code

A.1 Getting Started

R code for the book has been provided as online supplementary material and can be accessed at https://github.com/tpourmohamad/BayesianOptimizationBook. In order to be able to run the R code in the online supplement, a handful of packages from the Comprehensive R Archive Network (CRAN: https://cran.r-project.org/) must be first installed and then loaded. Keep in mind that, as of the writing of this book, thousands of packages exist on CRAN and so these are certainly not the only R packages that can be used for Gaussian process regression or Bayesian optimization, but rather are the packages that we prefer to work with for one reason or another.

To get started, R packages can be installed by use of the `install.packages()` command, where the name of the desired package is placed inside the function in quotes. For example, `install.packages("tgp")` installs the treed Gaussian process (`tgp`) package. Once installed, the `library()` command can then be used to load the package. The following is a list of R packages that we recommend you load before proceeding to any other code examples.

```
library(GA)            # Package for 3D plotting
library(tgp)           # Package for treed GPs
library(laGP)          # Package for fitting GPs
library(mvtnorm)       # Package for multivariate normal distribution
library(ggplot2)       # Package for plotting
library(gridExtra)     # Package for arranging multiple grid-based plots
library(CompModels)    # Package containing computer models
library(scatterplot3d) # Package for 3D scatter plots
```

We aimed at making all examples and figures in the book as reproducible as possible. However, even when setting seeds and using the same R version, one can get different random number streams on different architectures. Thus, we refrained from setting seeds in the R code with `set.seed()`. When running the R code you will invariably get different results than what are presented in the book, however, the

T. Pourmohamad, H. K. H. Lee, *Bayesian Optimization with Application to Computer Experiments*, SpringerBriefs in Statistics, https://doi.org/10.1007/978-3-030-82458-7

results should either be close to the original solutions and figures, or interpretable in a similar fashion.

By no means is R the only software language that exists for conducting Bayesian optimization. In fact, a significant amount of effort for developing Bayesian optimization software has been undertaken by the Python programming community. For example, the Python BoTorch software (Balandat et al. 2020) provides users with an out-of-the-box software for implementing Bayesian optimization with minimal effort.

Lastly, we note that all calculations and analyses in the book were carried out using R version 4.0.5 (2021-03-31)--"Shake and Throw".

References

Aarseth SJ, Hoyle F (1963) Dynamical evolution of clusters of galaxies, I. Mon Not R Astronom Soc 126(3):223–255

Ariafar S, Coll-Font J, Brooks D, Dy J (2019) ADMMBO: Bayesian optimization with unknown constraints using ADMM. J Mach Learn Res 20(123):1–26

Audet C, Kokkolaras M, Le Digabel S, Talgorn B (2018) Order-based error for managing ensembles of surrogates in mesh adaptive direct search. J Global Optim 70(3):645–675

Ba S, Joseph VR (2011) Multi-layer designs for computer experiments. J Am Stat Assoc 106:1139–1149

Baker E, Barbillon P, Fadikar A, Gramacy RB, Herbei R, Higdon D, Huang J, Johnson LR, Ma P, Mondal A, Pires B, Sacks J, Sokolov V (2021) Analyzing stochastic computer models: a review with opportunities. Stat Sci (to appear)

Balandat M, Karrer B, Jiang DR, Daulton S, Letham B, Wilson AG, Bakshy E (2020) BoTorch: a framework for efficient Monte-Carlo Bayesian optimization. In: Advances in neural information processing systems, vol 33. http://arxiv.org/abs/1910.06403

Banerjee S, Gelfand AE, Carlin BP (2004) Hierarchical Modeling and Analysis for Spatial Data. Chapman & Hall/CRC, New York

Bebber DV, Hochkirchen T, Siebertz K (2010) Statistische Versuchsplanung: Design of Experiments (DoE), 1st edn. Springer, Berlin, Heidelberg

Berger JO (1985) Statistical decision theory and Bayesian analysis, 2nd edn. Springer series in statistics. Springer, New York

Berger JO, De Oliveira V, Sansó B (2000) Objective Bayesian analysis of spatially correlated data. J Am Stat Assoc 96:1361–1374

Bertsekas DP (1982) Constrained optimization and Lagrange multiplier methods. Academic, New York

Binois M, Ginsbourger D, Roustant O (2015) A warped kernel improving robustness in bayesian optimization via random embeddings. In: Dhaenens C, Jourdan L, Marmion ME (eds) Learning and intelligent optimization. Springer International Publishing, Cham, pp 281–286

Binois M, Ginsbourger D, Roustant O (2020) On the choice of the low-dimensional domain for global optimization via random embeddings. J Global Optim 76(1):69–90

Bishop CM (1995) Neural networks for pattern recognition. Oxford University Press, New York

Breiman L (2001) Random forests. Mach Learn 45:5–32

Brochu E, Cora VM, de Freitas N (2010) A tutorial on bayesian optimization of expensive cost functions, with application to active user modeling and hierarchical reinforcement learning. Tech. Rep. 1012.2599, arXiv

© The Author(s), under exclusive license to Springer Nature Switzerland AG 2021
T. Pourmohamad, H. K. H. Lee, *Bayesian Optimization with Application to Computer Experiments*, SpringerBriefs in Statistics, https://doi.org/10.1007/978-3-030-82458-7

Broyden CG (1970) The convergence of a class of double-rank minimization algorithms. IMA J Appl Math 6(1):76–90

Chipman H, George E, McCulloch R (1998) Bayesian CART model search (with discussion). J Am Stat Assoc 93:935–960

Chipman H, George E, McCulloch R (2002) Bayesian treed models. Mach Learn 48:303–324

Chipman HA, George EI, McCulloch RE (2010) BART: Bayesian additive regression trees. Ann Appl Stat 4(1):266–298

Cox DD, John S (1997) Sdo: A statistical method for global optimization. In: Alexandrov NM, Hussaini MY (eds) Multidisciplinary Design Optimization: State of the Art. SIAM, Philadelphia, pp 315–329

Cunningham JP, Hennig P, Lacoste-Julien S (2013) Gaussian probabilities and expectation propagation. arXiv:1111.6832

Efstathiou G, Eastwood JW (1981) On the clustering of particles in an expanding Universe. Mon Not R Astron Soc 194:503–525

Eriksson D, Poloczek M (2021) Scalable constrained bayesian optimization. In: Banerjee A, Fukumizu K (eds) Proceedings of the 24th international conference on artificial intelligence and statistics, PMLR. Proceedings of machine learning research, vol 130, pp 730–738

Eriksson D, Dong K, Lee E, Bindel D, Wilson AG (2018) Scaling gaussian process regression with derivatives. In: Bengio S, Wallach H, Larochelle H, Grauman K, Cesa-Bianchi N, Garnett R (eds) Advances in neural information processing systems, vol 31. Curran Associates, Inc., Red Hook, pp 6867–6877

Eriksson D, Pearce M, Gardner J, Turner RD, Poloczek M (2019) Scalable global optimization via local bayesian optimization. In: Wallach H, Larochelle H, Beygelzimer A, d' Alché-Buc F, Fox E, Garnett R (eds) Advances in neural information processing systems, vol 32. Curran Associates, Inc., Red Hook, pp 5497–5508

Fang KT (1980) The uniform design: application of number-theoretic methods in experimental design. Acta Math Appl Sin 3:363–372

Fang KT, Li R, Sudjianto A (2005) Design and modeling for computer experiments. Chapman & Hall/CRC, Boca Raton

Fitzpatrick R (2012) An introduction to celestial mechanics. Cambridge University Press, Cambridge

Fletcher R (1970) A new approach to variable metric algorithms. Comput J 13(3):317–322

Gardner JR, Kusner MJ, Xu Z, Weinberger KQ, Cunningham JP (2014) Bayesian optimization with inequality constraints. In: Proceedings of the 24th international conference on machine learning, ICML '14, pp 937–945

Gill J (2014) Bayesian methods: a social and behavioral sciences approach, 3rd edn. Chapman and Hall/CRC, Boca Raton

Goldfarb DF (1970) A family of variable-methods derived by variational means. Math Comput 24:23–26

Gonzalez J, Dai Z, Hennig P, Lawrence N (2016) Batch Bayesian optimization via local penalization. In: Gretton A, Robert CC (eds) Proceedings of the 19th international conference on artificial intelligence and statistics, Cadiz. Proceedings of machine learning research, vol 51, pp 648–657

Gramacy RB (2007) TGP: an R package for bayesian nonstationary, semiparametric nonlinear regression and design by treed Gaussian process models. J Stat Softw 19(9):1–46

Gramacy RB (2020) Surrogates: Gaussian process modeling, design, and optimization for the applied sciences. Chapman & Hall/CRC, Boca Raton

Gramacy RB, Lee HKH (2008) Bayesian treed Gaussian process models with an application to computer modeling. J Am Stat Assoc 103:1119–1130

Gramacy RB, Lee HKH (2011) Optimization under unknown constraints. In: Bernardo J, Bayarri S, Berger J, Dawid A, Heckerman D, Smith A, West M (eds) Bayesian statistics, vol 9. Oxford University Press, Oxford, pp 229–256

Gramacy RB, Lee HKH (2012) Cases for the nugget in modeling computer experiments. Stat Comput 22:713–722

Gramacy RB, Taddy M (2010) Categorical inputs, sensitivity analysis, optimization and importance tempering with TGP version 2, an R package for treed Gaussian process models. J Stat Softw 33(6):1–48

Gramacy RB, Gray GA, Le Digabel S, Lee HKH, Ranjan P, Wells G, Wild SM (2016) Modeling an augmented lagrangian for blackbox constrained optimization. Technometrics 58(1):1–11

Hardy RL (1971) Multiquadratic equations of topography and other irregular domains. J Geophys Res 76(8):1905–1915

Henning P, Schuler CJ (2012) Entropy search for information-efficient global optimization. J Mach Learn Res 13:1809–1837

Hernández-Lobato JM, Hoffman MW, Ghahramani Z (2014) Predictive entropy search for efficient global optimization of black-box functions. In: Ghahramani Z, Welling M, Cortes C, Lawrence N, Weinberger K (eds) Advances in neural information processing systems, vol 27. Curran Associates, Red Hook

Hernández-Lobato JM, Requeima J, Pyzer-Knapp EO, Aspuru-Guzik A (2017) Parallel and distributed thompson sampling for large-scale accelerated exploration of chemical space. In: Precup D, Teh YW (eds) Proceedings of the 34th international conference on machine learning. Proceedings of machine learning research, vol 70, pp 1470–1479

Higdon D (1998) A process-convolution approach to modeling temperatures in the North Atlantic Ocean. J Environ Ecol Stat 5(2):173–190

Higdon D (2002) Space and space–time modeling using process convolutions. In: Anderson C, Barnett V, Chatwin PC, El-Shaarawi AH (eds) Quantitative methods for current environmental issues. Springer, London, pp 37–56

Hoff PD (2009) A first course in Bayesian statistical methods. Springer, New York, NY

Jalali H, Van Nieuwenhuyse I, Picheny V (2017) Comparison of kriging-based methods for simulation optimization with heterogeneous noise. Eur J Oper Res 261(1):279–301

Johnson ME, Moore LM, Ylvisaker D (1990) Minimax and maximin distance designs. J Stat Plann Infer 26:131–148

Jones DR, Schonlau M, Welch WJ (1998) Efficient global optimization of expensive black box functions. J Global Optim 13:455–492

Jones B, Silvestrini RT, Montgomery DC, Steinberg DM (2015) Bridge designs for modeling systems with low noise. Technometrics 57:155–163

Joseph VR (2016) Space-filling designs for computer experiments: a review. Qual Eng 28(1):28–35

Joseph VR, Dasgupta T, Tuo R, Wu CFJ (2015) Sequential exploration of complex surfaces using minimum energy designs. Technometrics 57:64–74

Kandasamy K, Krishnamurthy A, Schneider J, Poczos B (2018) Parallelised bayesian optimisation via thompson sampling. In: Storkey A, Perez-Cruz F (eds) Proceedings of the twenty-first international conference on artificial intelligence and statistics. Proceedings of machine learning research, vol 84, pp 133–142

Kennedy M, O'Hagan A (2001) Bayesian calibration of computer models. J R Stat Soc Ser B (Stat Methodol) 63(2):425–464

Kirschner J, Mutny M, Hiller N, Ischebeck R, Krause A (2019) Adaptive and safe Bayesian optimization in high dimensions via one-dimensional subspaces. In: Chaudhuri K, Salakhutdinov R (eds) Proceedings of the 36th international conference on machine learning, PMLR. Proceedings of machine learning research, vol 97, pp 3429–3438

Kleijnen JPC (2015) Design and analysis of simulation experiments, 2nd edn. Springer, New York

Klypin AA, Trujillo-Gomez S, Primack J (2011) Dark matter halos in the standard cosmological model: results from the Bolshoi simulation. Astrophys J 740(2):1–17

Kushner HJ (1964) A new method of locating the maximum of an arbitrary multipeak curve in the presence of noise. J Basic Eng 86:97–106

Lawrence E, Heitmann K, Kwan J, Upadhye A, Bingham D, Habib S, Higdon D, Pope A, Finkel H, Frontiere N (2017) The mira-titan universe. II. Matter power spectrum emulation. Astrophys J 847(1):50. https://doi.org/10.3847/1538-4357/aa86a9

Lee HKH, Gramacy RB, Linkletter C, Gray GA (2011) Optimization subject to hidden constraints via statistical emulation. Pacif J Optim 7:467–478

Letham B, Karrer B, Ottoni G, Bakshy E (2019) Constrained Bayesian optimization with noisy experiments. Bayesian Anal 14(2):495–519

Letham B, Calandra R, Rai A, Bakshy E (2020) Re-examining linear embeddings for high-dimensional bayesian optimization. In: Larochelle H, Ranzato M, Hadsell R, Balcan MF, Lin H (eds) Advances in neural information processing systems, vol 33. Curran Associates, Inc., Red Hook, pp 1546–1558

Lindberg D, Lee HKH (2015) Optimization under constraints by applying an asymmetric entropy measure. J Comput Graph Stat 24:379–393

Marcellin S, Zighed DA, Ritschard G (2006) An asymmetric entropy measure for decision trees. In: 11th information processing and management of uncertainty in knowledge-based systems (IPMU 06), pp 1292–1299

Mathesen L, Pedrielli G, Ng SH, Zabinsky ZB (2021) Stochastic optimization with adaptive restart: a framework for integrated local and global learning. J Global Optim 79:87–110

Matott LS, Leung K, Sim J (2011) Application of matlab and python optimizers to two case studies involving groundwater flow and contaminant transport modeling. Comput Geosci 37:1894–1899

Mayer AS, Kelley CT, Miller CT (2002) Optimal design for problems involving flow and transport phenomena in saturated subsurface systems. Adv Water Resour 25:1233–1256

McDonald MG, Harbaugh AW (2003) The history of MODFLOW. Ground Water 41(2):280–283

McKay MD, Conover WJ, Beckman RJ (1979) A comparison of three methods for selecting values of input variables in the analysis of output from a computer code. Technometrics 21:239–245

Meng X, Zhang H, Mezei M, Cui M (2011) Molecular docking: a powerful approach for structure-based drug discovery. Curr Comput-Aided Drug Des 7(2):239–245

Mockus J, Tiesis V, Zilinskas A (1978) The application of Bayesian methods for seeking the extrenum. Towards Global Optim 2:117–129

Moustaki I, Knott M (2000) Generalized latent trait models. Psychometrika 65:391–411

Mutny M, Krause A (2018) Efficient high dimensional bayesian optimization with additivity and quadrature fourier features. In: Bengio S, Wallach H, Larochelle H, Grauman K, Cesa-Bianchi N, Garnett R (eds) Advances in neural information processing systems, vol 31. Curran Associates, Inc., Red Hook, pp 9005–9016

Nayebi A, Munteanu A, Poloczek M (2019) A framework for Bayesian optimization in embedded subspaces. In: Chaudhuri K, Salakhutdinov R (eds) Proceedings of the 36th international conference on machine learning, PMLR. Proceedings of machine learning research, vol 97, pp 4752–4761

Negoescu DM, Frazier PI, Powell WB (2011) The knowledge-gradient algorithm for sequencing experiments in drug discovery. INFORMS J Comput 23(3):346–363

Nelder J, Mead R (1965) A simplex method for function optimization. Comput J 7(4):308–313

Nocedal J, Wright SJ (2006) Numerical optimization, 2nd edn. Springer, New York, NY

Oh C, Gavves E, Welling M (2018) BOCK: Bayesian optimization with cylindrical kernels. In: Proceedings of the 35th international conference on machine learning, ICML 2018, vol 80, pp 3868–3877

Picheny V (2014) A stepwise uncertainty reduction approach to constrained global optimization. Proceedings of the seventeenth international conference on artificial intelligence and statistics, pp 787–795

Picheny V, Ginsbourger D (2014) Noisy kriging-based optimization methods: a unified implementation within the diceoptim package. Comput Stat Data Anal 71:1035–1053

Picheny V, Wagner T, Ginsbourger D (2013) A benchmark of kriging-based infill criteria for noisy optimization. Struct Multidiscipl Optim 48(3):607–626

Picheny V, Gramacy RB, Wild S, Le Digabel S (2016) Bayesian optimization under mixed constraints with a slack-variable augmented Lagrangian. In: Lee D, Sugiyama M, Luxburg U, Guyon I, Garnett R (eds) Advances in neural information processing systems, vol 29. Curran Associates, Inc., Red Hook, pp 1435–1443

Pourmohamad T (2020) CompModels: Pseudo Computer Models for Optimization. https://CRAN.R-project.org/package=CompModels. r package version 0.2.0

Pourmohamad T, Lee HKH (2016) Multivariate stochastic process models for correlated responses of mixed type. Bayesian Anal 11(3):797–820

Pourmohamad T, Lee HKH (2020) The statistical filter approach to constrained optimization. Technometrics 62(3):303–312

Pourmohamad T, Lee HKH (2021) Bayesian optimization via barrier functions. J Comput Graph Stat (to appear). https://www.tandfonline.com/doi/full/10.1080/10618600.2021.1935270

Qian PZG, Wu CFJ (2009) Sliced space-filling designs. Biometrika 96:945–956

Qian PZG, Tang B, Wu CFJ (2009) Nested space-filling designs for experiments with two levels of accuracy. Stat Sin 19:287–300

Quevauviller P (ed) (2007) Groundwater science and policy: an international overview. Royal Society of Chemistry, Cambridge

Quiroga R, Villarreal MA (2016) Vinardo: a scoring function based on Autodock Vina improves scoring, docking, and virtual screening. PLoS One 11(5):1231–1249

R Core Team (2021) R: a language and environment for statistical computing. R Foundation for Statistical Computing, Vienna. https://www.R-project.org/

Rasmussen CE, Williams CKI (2006) Gaussian processes for machine learning. The MIT Press, Cambridge, MA

Rogers K, Peiris H, Pontzen A, Bird S, Verde L, Font-Ribera A (2019) Bayesian emulator optimisation for cosmology: application to the Lyman-alpha forest. J Cosmol Astropart Phys 2019(2):031–031

Rolland P, Scarlett J, Bogunovic I, Cevher V (2018) High-dimensional Bayesian optimization via additive models with overlapping groups. Proceedings of the twenty-first international conference on artificial intelligence and statistics, pp 298–307

Sacks J, Welch WJ, Mitchell TJ, Wynn HP (1989) Design and analysis of computer experiments. Stat Sci 4:409–435

Sammel MD, Ryan LM, Legler JM (1997) Latent variable models for mixed discrete and continuous outcomes. J R Stat Soc Ser B (Stat Methodol) 59:667–678

Santner TJ, Williams BJ, Notz WI (2003) The design and analysis of computer experiments. Springer, New York, NY

Schonlau M, Welch W, Jones D (1998) Global versus local search in constrained optimization of computer models. In: New Developments and applications in experimental design, vol 34. IMS lecture notes. Institute of mathematical statistics, pp 11–25

Shahriari B, Swersky K, Wang Z, Adams RP, de Freitas N (2016) Taking the human out of the loop: A review of Bayesian optimization. Proc IEEE 104(1):148–175

Shanno D (1970) Conditioning of quasi-Newton methods for function minimization. Math Comput 24:647–656

Shewry MC, Wynn HP (1987) Maximum entropy sampling. J Appl Stat 14:165–170

Srinivas N, Krause A, Kakade S, Seeger M (2010) Gaussian process optimization in the bandit setting: No regret and experimental design. In: Proceedings of the 27th international conference on machine learning, ICML 2010

Stein ML (1999) Interpolation of spatial data. Springer, New York, NY

Svenson JD, Santner TJ (2016) Multiobjective optimization of expensive-to-evaluate deterministic computer simulator models. Comput Stat Data Anal 94:250–264

Thompson WR (1933) On the likelihood that one unknown probability exceeds another in view of the evidence of two samples. Biometrika 25(3/4):285–294

Trott O, Olson J (2010) Autodock Vina: improving the speed and accuracy of docking with a new scoring function, efficient optimization and multithreading. J Comput Chem 31:455–461

Wackernagel H (2003) Multivariate geostatistics: an introduction with applications. Springer, New York, NY

Wang Z, Hutter F, Zoghi M, Matheson D, de Freitas N (2016) Bayesian optimization in a billion dimensions via random embeddings. J Artif Intell Res 55:361–387

Wang Z, Gehring C, Kohli P, Jegelka S (2018) Batched large-scale Bayesian optimization in high dimensional spaces. Proceedings of the twenty-first international conference on artificial intelligence and statistics, pp 745–754

Zhang B, Cole DA, Gramacy RB (2021) Distance-distributed design for Gaussian process surrogates. Technometrics 63(1):40–52

Printed in the United States
by Baker & Taylor Publisher Services